Climate Impacts on the Baltic Sea: From Science to Policy

Springer Earth System Sciences

For further volumes:
http://www.springer.com/series/10178

Marcus Reckermann • Keith Brander •
Brian R. MacKenzie • Anders Omstedt
Editors

Climate Impacts on the Baltic Sea: From Science to Policy

School of Environmental Research –
Organized by the Helmholtz-Zentrum
Geesthacht

Springer

Editors
Marcus Reckermann
Helmholtz-Zentrum Geesthacht
(HZG)
Max-Planck-Strasse 1
Geesthacht
Germany

Keith Brander
Technical University of Denmark
Jaegersborg Allé 1
Charlottenlund
Denmark

Brian R. MacKenzie
Technical University of Denmark
Jaegersborg Allé 1
Charlottenlund
Denmark

Anders Omstedt
University of Gothenburg
Earth Sciences Centre
Gothenburg
Sweden

ISBN 978-3-642-25727-8 ISBN 978-3-642-25728-5 (eBook)
DOI 10.1007/978-3-642-25728-5
Springer Heidelberg New York Dordrecht London

Library of Congress Control Number: 2012936488

Printed on acid-free paper

Springer is part of Springer Science+Business Media (www.springer.com)

Preface

The Baltic Sea area is an old cultural landscape with a well developed international framework for monitoring, assessing and managing its marine ecosystems. It provides a good case study for other regions where such management is being set up. One of our main aims is to introduce this example of management in action to a wider community. The existing Baltic Sea Action Plan, which is a major component of the management framework, has not yet been extended to include climate change impacts, so another aim is to supply some of the background needed to do this.

With a catchment basin covering almost 20% of the European continent, the Baltic Sea is one of the largest brackish seas in the world. It spans different climate and population zones, from the temperate, highly populated and industrialized south with intensive agriculture, to the boreal and rural north. The Baltic Sea features some natural peculiarities which make its ecosystems especially vulnerable to external stresses due to pollution, eutrophication and climate change. Water exchange with the North Sea and world oceans is limited; the deep basins are separated by sills, and there is a steep salinity gradient from south to north. In the past decades, the Baltic Sea has demonstrated its sensitivity to various human pressures and to climate change.

This book is the outcome of a Summer School titled "Climate Impacts on the Baltic Sea: From Science to Policy", which took place in Nexø on the Danish Baltic Sea island of Bornholm, 27 July to 5 August 2009. The purpose of the summer school was to provide an introduction to the entire range of climate change impacts on the Baltic Sea, including its catchment basin, from the physical and biological to the societal and political impacts. An interdisciplinary approach was followed: students from different natural science disciplines coming together to develop skills in observing, modelling, projecting and interpreting physical and biological changes in the Baltic Sea. An important goal of the course was to demonstrate how discipline-specific scientific knowledge can be combined in order to contribute to real management solutions at the ecosystem level, and how scientific assessments of climate change impacts can be communicated to policy makers and the general public. Such interdisciplinary assessments are increasingly required as part of the management of societal pressure on marine ecosystems and to deal with the additional pressure from climate change.

Twenty students from all across the Baltic Sea came together in the "Møbelfabrikken", a former furniture factory, which proved to be a perfect venue for the course. Sunny, warm weather and occasional recreational swims at the nearby rocky beaches contributed to the good spirit. The facilities are close to 'Bornholms lakseklækkeri' (Bornholm Salmon Hatchery, owned by DTU-Aqua and used as an experimental cod hatchery and release facility for stock enhancement) which was also visited during the course.

Students and lecturers in front of the *Møbelfabrikken*

Students were assigned activities to develop their discipline-specific knowledge, learn how the knowledge could be translated to management and policy actions, and improve general scientific skills (e.g. oral and written presentation of scientific results; organisation and structuring of group work; collaboration with colleagues from different nations and disciplines). The course contained lectures, individual exercises, discussion tutorials and group exercises. One type of group exercise was discipline-specific (e.g. related to modelling oceanographic or fish population responses to climate change), while the second was course-long, interdisciplinary and oriented to management objectives for the Baltic Sea ecosystem. This overarching task was the preparation and presentation by four student groups of a "climate change action plan" that would supplement the existing HELCOM Baltic Sea Action Plan. The task inspired some long evening and night sessions among students and lecturers. At the end of the course, students were asked to give an oral presentation of their visions. The outcomes of this exercise (one page summaries and presentations) are available at the course website: www.baltex-research.eu/baltic2009.

The chapters in this book are based on lectures given at the course. They cover a range of topics, spanning from detailed descriptions of political agreements that protect the marine environment, to basic modelling instructions, to an assessment of the possible impacts of climate change on the marine ecosystem, to a reflection on the role of climate scientists and their responsibility in society.

Chris Hopkins presents an extensive overview over international environmental agreements and conventions and demonstrates their role, importance and effectiveness for the environmental recovery of the Baltic Sea. This chapter represents an unprecedented summary of these activities in the Baltic Sea region and Europe in general (Chap. 1). **Minna Pyhälä** focuses on a recent and important activity: the HELCOM Baltic Sea Action Plan (BSAP) as the prominent international agreement aiming at a fundamental recovery of the Baltic Sea (Chap. 2). **Anders Omstedt** discusses basic oceanographic features of the Baltic Sea, and gives a basic introduction how these processes, including important aspects of the carbon cycle, can be modelled (Chap. 3). **Eduardo Zorita** provides a basic introduction to the principles of climate modelling with a special emphasis on the related uncertainties (Chap. 4). **Keith Brander** and **Jon Havenhand** give a short overview of the observed and possible future impacts of climate change on the marine ecosystems, including acidification (Chap. 5), and **Rabea Diekmann** and co-authors present some background on ecosystem theory related to abrupt ecosystem changes referred to as regime shifts, including a description of statistical techniques which can be used to analyze such regime shifts (Chap. 6). Sustainability is a central term in the protection and management of ecosystems, but it is also essential that scientists themselves act in a responsible and sustainable way, as explained by **Hans von Storch** (Chap. 7).

This interdisciplinary book is primarily directed at students and lecturers of the environmental disciplines to provide an overview of the possible impacts of climate change on the Baltic Sea. We also intend it to serve as a good background reference for scientists and policy makers, both for the Baltic Sea area and more generally.

Marcus Reckermann
Keith Brander
Brian MacKenzie
Anders Omstedt

Acknowledgments

The summer school was co-organized by the Danish Technical University DTU-Aqua, the University of Gothenburg and BALTEX, the interdisciplinary research network for the Baltic Sea basin. The school was primarily supported by the Nordic Marine Academy (NMA) as an Advanced Course, and by BONUS, the Baltic Organisations Network for Funding Science EEIG. Further support came from the Danish Technical University (DTU-Aqua), the University of Gothenburg, Helmholtz-Zentrum Geesthacht (HZG, formerly GKSS) and its "Gesellschaft zur Förderung des GKSS Forschungszentrums e.V.", fishnet.dk, EUR-OCEANS and BALTEX. The summer school is a contribution to the BONUS+ projects Baltic-C and ECOSUPPORT.

The idea of a summer school initiated within the BALTEX and the EUR-OCEANS programmes. The editors and course organizers would like to thank Fritz Köster and David Turner for early discussions on the possibility to organize a joint EUR-OCEANS-BALTEX summer school on Bornholm, and Götz Flöser of Helmholtz-Zentrum Geesthacht (HZG) for the opportunity to publish this book in a book series of HZG summer schools. The reviewers of chapter manuscripts are particularly thanked for their valuable comments to improve the quality of the manuscripts. Thank you also to Silke Köppen at the International BALTEX Secretariat who put much effort into the technical editing of the manuscripts and obtaining permissions. Furthermore, we would like to thank all lecturers for their engagement and last but not least the students for their infectious enthusiasm, which made this summer school such a pleasant and valuable experience.

The editors and course organizers, July 2011

Marcus Reckermann	International BALTEX Secretariat at Helmholtz-Zentrum Geesthacht, Germany
Keith Brander	Danish Technical University, National Institute of Aquatic Resources (DTU Aqua), Charlottenlund, Denmark
Brian MacKenzie	Danish Technical University, National Institute of Aquatic Resources (DTU Aqua), Charlottenlund, Denmark
Anders Omstedt	Department of Earth Sciences, University of Gothenburg, Sweden

From left to right: Marcus Reckermann, Anders Omstedt, Keith Brander and Brian MacKenzie

nma
Nordic Marine Academy

BONUS
Baltic Organisations Network for Funding Science EEIG

DTU

BALTEX

Helmholtz-Zentrum Geesthacht
Centre for Materials and Coastal Research

Gesellschaft zur Förderung des GKSS-Forschungszentrums e.V.

fishnet.dk

Oceans

Contents

Contributors

Keith Brander Technical University of Denmark, Charlottenlund, Denmark, kbr@aqua.dtu.dk

Rabea Diekmann Institute for Hydrobiology and Fisheries Science, University of Hamburg, Hamburg, Germany, rabea.diekmann@uni-hamburg.de

Jon Havenhand Department of Marine Ecology – Tjärnö, University of Gothenburg, Strömstad, Sweden, jon.havenhand@marecol.gu.se

Christopher C.E. Hopkins AquaMarine Advisers, Åstorp, Sweden, chris.hopkins@aquamarine.se

Brian R. MacKenzie Technical University of Denmark, Charlottenlund, Denmark, brm@aqua.dtu.dk

Christian Möllmann Institute for Hydrobiology and Fisheries Science, University of Hamburg, Hamburg, Germany, christian.moellmann@uni-hamburg.de

Anders Omstedt Department of Earth Sciences, University of Gothenburg, Gothenburg, Sweden, Anders.Omstedt@gvc.gu.se

Saskia Otto Institute for Hydrobiology and Fisheries Science, University of Hamburg, Hamburg, Germany, saskia.otto@uni-hamburg.de

Minna Pyhälä Secretariat of the Helsinki Commission, Helsinki, Finland, minna.pyhala@helcom.fi

Marcus Reckermann International BALTEX Secretariat, Helmholtz-Zentrum Geesthacht, Geesthacht, Germany, marcus.reckermann@hzg.de

Hans von Storch Institute of Coastal Research, Helmholtz-Zentrum Geesthacht, Geesthacht, Germany; Hamburg University, Hamburg, Germany, hvonstorch@web.de

Eduardo Zorita Institute of Coastal Research, Helmholtz-Zentrum Geesthacht, Geesthacht, Germany, eduardo.zorita@hzg.de

Chapter 1
International Agreements and Baltic Sea Environmental Management

Christopher C.E. Hopkins

Abbreviations

ASCOBANS	Agreement on the Conservation of Small Cetaceans of the Baltic and North Seas
BAT	Best Available Technology
BEP	Best Environmental Practice
BONUS	Baltic Organisations Network for Funding Science EEIG
BS	Baltic Sea
BSAP	Baltic Sea Action Plan (HELCOM)
BSPA	Baltic Sea Protected Area (HELCOM)
CBD	Convention on Biological Diversity
CFP	Common Fisheries Policy (EU)
CITES	Convention on International Trade in Endangered Species of Wild Fauna and Flora
COP	Conference of the Parties
DPSIR	Drivers-Pressures-Status-Impacts-Responses
EAM	Ecosystem Approach to Management of Human Activities
EC	European Community
EcoQO	Ecological Quality Objective
EEC	European Economic Community (predecessor to EC and EU)
EEIG	European Economic Interest Grouping
EEZ	Exclusive Economic Zone
EIA	Environmental Impact Assessment
EU	European Union
FAO	UN Food and Agriculture Organization
GES	Good Environmental Status
GEcS	Good Ecological Status
GHG	Greenhouse gas

C.C.E. Hopkins (✉)
AquaMarine Advisers, Granvägen 20, SE-265 32 Åstorp, Sweden
e-mail: chris.hopkins@aquamarine.se

M. Reckermann et al. (eds.), *Climate Impacts on the Baltic Sea: From Science to Policy*, Springer Earth System Sciences, DOI 10.1007/978-3-642-25728-5_1,

HABITAT	Nature Protection and Biodiversity Group (HELCOM)
HELCOM	Helsinki Commission – Baltic Marine Environmental Commission
IBSFC	International Baltic Sea Fishery Commission (disbanded)
ICES	International Council for the Exploration of the Sea
IMO	International Maritime Organization (UN)
IPCC	Intergovernmental Panel on Climate Change
JMM	Joint Ministerial Meeting of HELCOM and OSPAR
LME	Large Marine Ecosystem
LRP	Limit Reference Point
MARITIME	Maritime Group (HELCOM)
MARPOL	International Convention for the Prevention of Pollution from Ships, 1973 as modified by the Protocol of 1978
MONAS	Monitoring and Assessment Group (HELCOM)
MPA	Marine Protected Area
MSFD	Marine Strategy Framework Directive
MSY	Maximum Sustainable Yield
NGO	Non-Governmental Organization
NOAA	US National Oceanic and Atmospheric Administration
NSC	North Sea Conference
OSPAR	OSPAR Commission for the Protection of the Marine Environment of the North-East Atlantic
PP	Precautionary Principle
PPP	Polluter-Pays Principle
RAC	Regional Advisory Council (for EU fisheries)
RESPONSE	Response Group (HELCOM)
RTD	Research and Technological Development
SAC	Special Area of Conservation
SEA	Strategic Environmental Impact Assessment
SPA	Special Protection Area
UN	United Nations
UNCCD	UN Convention to Combat Desertification
UNCED	UN Conference on Environment and Development
UNCLOS	United Nations Convention on the Law Of the Sea
UNECE	UN Economic Commission for Europe
UNEP	UN Environment Programme
UNFCCC	UN Framework Convention on Climate Change
WFD	Water Framework Directive
WSSD	World Summit for Sustainable Development

1.1 Introduction

Climate change caused by the emission of global warming gases due to human activities is potentially the most significant threat to the natural environment and biodiversity, and to the sustainable use of biodiversity by humans, at both the global and regional scales (IPCC 2001; CBD 2003; HELCOM 2007a; Jennings and Brander 2010). The regionalization of climate variability and climate change is a major goal of research and management in shelf seas such as the Baltic Sea.

The Intergovernmental Panel on Climate Change (IPCC) has increasingly recognized that the impacts of climate are overlaid on the impacts of other pressures (e.g. excessive 'extractive' harvesting including overfishing, pollution including eutrophication, and land and habitat degradation, and introductions of invasive alien species), and that these multiple pressures frequently have synergistic, detrimental effects (IPCC 2007a, b; see also Perry et al. 2010 and Planque et al. 2010 for climate – fisheries interactions). Thus, climate change cannot be seen as an isolated phenomenon but rather as part of a series of impacts caused by unsustainable practices, with linkages to several other environmental challenges. So, management, and the scientific knowledge that it depends on, must adopt a system-level perspective that considers climate patterns, trends and episodes in the context of these other, multiple stressors. However, until recently, most environmental assessments of regional seas, including the Baltic Sea region, have not given sufficient attention to identifying and understanding environmental forcing and especially the effects of climate change and variability on ecosystems (cf. Hopkins 2004; HELCOM 2003, 2007a, 2009, 2010). Additionally, few of the many major marine environmental policy and management conventions, agreements and other instruments specifically address, in their basic texts, the need to address climate change. There is, however, an increasing appreciation that climate change must be practically integrated into their work-plans and operations.

Addressing climate change requires two types of response. Firstly, and importantly, we must reduce our greenhouse gas (GHG) emissions (i.e. take mitigatory action), and secondly, we must take adaptation action to deal with (i.e. learn to live with) the unavoidable impacts.

Climate change will increasingly impact, over the coming decades, upon the biological, economic and social aspects of ecosystems (IPCC 2007a, b). As climate change impacts may vary between detrimental and beneficial depending on the regional environment, it will pose challenges and provide opportunities. It is important to identify, predict, mitigate and adapt to the scale and magnitude of the change acting on the ecosystem and the dependent human activities. In this paper it is argued that, as effects of climate forcing also depend on the degree to which other pressures are also causing stress, the aim of achieving good environmental status of Europe's marine environment would create a desirable 'buffer' limiting the malevolence of the unwelcome impacts.

1.2 1992 'Rio Earth Summit': Parent of the 'Ecosystem Approach' and Climate Change Mitigation

The Brundtland Commission, convened by the United Nations (UN) in 1983, in its 1987 report (commonly called the Brundtland report) (Brundtland 1987) alerted the world to the urgency of making progress toward economic development that could be sustained without depleting natural resources or harming the environment. In the report, sustainability (i.e. sustainable development) was defined as 'development that meets the needs of the present without compromising the ability of future generations to meet their own needs.'

The Brundtland report triggered the emergence of a raft of international conferences, agreements, conventions, and other instruments which further enshrined the concept of environmentally sustainable development. In particular, the 1992 UN Conference on Environment and Development (UNCED 1992), also called the Rio 'Earth Summit', underlined that the economic foundation for human wealth and health requires to be carefully integrated within a framework of ecological conservation and protection.

1.2.1 1992 UN Conference on Environment and Development (Rio 'Earth Summit')

The 1992 Rio 'Earth Summit' (UNCED 1992) adopted three ground-breaking international conventions: the UN Framework Convention on Climate Change (UNFCCC 1992), the UN Convention on Biological Diversity (CBD 1992), and the UN Convention to Combat Desertification (UNCCD 1994). These conventions are basically interlinked in that they, for the first time, inherently recognized the need to apply an 'ecosystem approach' in which humans are acknowledged as forming part of ecosystems and that human activities, including the production of pollutants and unsustainable utilization and degradation of ecosystem resources, need to be managed/regulated in order to achieve sustainable development. It was recognized that humans are, for example, causing harmful climate change as evidenced in global warming and expansion of desert areas. The 'precautionary principle' and the 'polluter-pays principle' (see Sect. 1.4.1) formed key components arising from UNCED and the subsequent UN Agenda 21[1] programme promoting sustainable development.

A decade later, at the 2002 World Summit on Sustainable Development (WSSD, also called Earth Summit 2002) which affirmed UN commitment to 'full implementation' of

---

[1] Agenda 21, referring to the twenty-first century, is a comprehensive blueprint of action to be taken globally, nationally and locally by UN organizations, governments, and major groups in every area in which humans impact on the environment.

Agenda 21, a significant milestone was reached for a Johannesburg Plan of Implementation that agreed on several specific ecosystem-related targets of relevance to the marine environment (UN 2002). These included: achievement of 'substantial' reductions in land-based sources of pollution by 2006; introduction of the ecosystems approach to marine resource assessment and management by 2010; designation of a network of marine protected areas by 2012; and the maintenance and restoration of fish stocks to maximum sustainable yield (MSY) levels by 2015.

1.2.2 1992 UN Framework Convention on Climate Change

The UNFCCC is the framework for the international negotiations on reducing GHG emissions. The UNFCCC aims to stabilize GHG concentrations 'at a level that would prevent dangerous anthropogenic (human induced) interference with the climate system.' *This level* 'should be achieved within a time-frame sufficient to allow ecosystems to adapt naturally to climate change, to ensure that food production is not threatened, and to enable economic development to proceed in a sustainable manner.' Thus, the convention clearly accepts the concept of sustainable, ecosystem-based management. The UNFCCC's Contracting Parties committed to cooperate concerning the conservation of sinks and reservoirs of GHGs such as the oceans and coastal and marine ecosystems.

The Kyoto Protocol (UN 1997) was drawn up in 1997 to implement the UNFCCC. The protocol, which entered into force in February 2005, committed the world's industrialized countries to undertake jointly in reducing GHG emissions by just over 5% between 2008 and 2012, compared with emissions levels in 1990. All the nine coastal Baltic Sea countries (eight EU and the Russian Federation) have ratified the agreement.

The UNFCCC-Kyoto approach, as articulated in the 2007 Bali Conference (UN 2007), can be summarized into three simple propositions: (1) developed countries should adopt national emissions reduction targets; (2) developing countries should undertake mitigation actions; and (3) developed countries should provide developing countries with financing and technological support for mitigation.

The Kyoto Protocol does not reduce emissions far enough, but it can be seen as a first important step towards the necessary substantial reduction. The EU, like many industrialized countries, is implementing extensive measures in order to meet its Kyoto commitments. In the international climate negotiations, the EU is pressing for legally binding emissions targets for industrialized countries. The new, post-Kyoto agreement should aim to reduce industrialized countries' emissions by 60–80% by 2050 and by 25–40% by 2020, compared to levels in 1990.

The Kyoto Protocol commitment period comes to an end in 2012. Currently, efforts are focused on finding a replacement regime that will determine the future basis upon which global cooperation on *reducing GHC emissions* (i.e. mitigation) and *adapting to climate change impacts* will take place. The December 2009 UN

Climate Conference COP 15 in Copenhagen (Denmark) (UN 2009) aimed to conclude negotiations on the new climate treaty to succeed the Kyoto Protocol, but failed to achieve consensus on the specific measures regarding reduction targets for emissions and the level of financial support for developing countries.

1.3 Transboundary and International 'Instruments' for Environmental Management of Coastal and Offshore Areas

The following sub-sections provide an overview of the main operative and developing management systems contributing to environmental management in the Baltic Sea region. These systems have the *potential*, subject to the political will to implement them, to combat the causes of overexploitation, pollution, habitat degradation and human encroachment. These systems include various international conventions, agreements and other instruments, as well as the important precautionary principle and the ecosystem approach to management. The intention here is to focus on international initiatives rather than those falling under the auspices of a single coastal country, in order to limit the review to a convenient size and to focus on transboundary collaboration.

1.3.1 Examples of Major Environmental 'Instruments' Applicable in the Baltic Sea Region

Numerous international conventions, agreements and other instruments contribute to the protection of the coastal and offshore environment and biodiversity of the Baltic Sea region (Table 1.1).

In 2004, Estonia, Latvia, Lithuania and Poland acceded to the European Union (EU), joining Denmark, Finland, Germany and Sweden. Accordingly, many European Economic Community/European Union[2] directives apply to the marine environment, and living marine resources including fisheries and biodiversity, in the national waters of these EU countries (Table 1.2). Although the various instruments may appear fragmented, they are all in the process of being subsumed by the *Ecosystem approach to the management of human activities* (EAM, see Sects. 1.4.2 and 1.4.3) across several areas of policy including pertinent directives.

[2] The European Economic Community (EEC) was renamed as the European Community (EC) in 1993. The succession of the EC by the European Union (EU) occurred on 1 December due to entering into force of the Treaty of Lisbon on that date.

Table 1.1 International conventions, agreements and codes of conduct/guidelines concerning the environment and biodiversity, including conservation of fishery resources, of the Baltic Sea region

Year	Instrument
1964	Convention for the International Council for the Exploration of the Sea (ICES, established in 1902), applying to the North Atlantic and its adjacent seas, including the Baltic Sea, aims to (a) promote and encourage research and investigations for the study of the marine environment and its living marine resources, (b) draw up programmes for this purpose and organize supporting research and investigations, (c) publish or otherwise disseminate the results of research and investigations. Since the 1970s, a major task is providing scientific information and advice to intergovernmental regulatory commissions (e.g. HELCOM, OSPAR), the European Commission, and ICES Member Countries, for fisheries conservation and protection of the marine environment and its associated biodiversity and ecosystems
1971	Ramsar Convention on Wetlands of International Importance, especially as Waterfowl Habitat
1972	Convention on Prevention of Marine Pollution by Dumping of Wastes and Other Matter (former London Dumping Convention), prohibits dumping at sea and bans disposal of radioactive waste at sea
1973	MARPOL 73/78 – IMO Convention on Marine Pollution from Ships. Limits operational discharges of oil, noxious liquids, and ship generated garbage including litter
1973	Convention on International Trade in Endangered Species of Wild Fauna and Flora (CITES) aims to ensure that international trade in specimens of wild animals and plants does not threaten their survival
1974/92	The Convention for the Protection of the Marine Environment of the Baltic Sea Area, established the Helsinki Commission (Baltic Marine Environment Protection Commission, HELCOM) in order to protect the marine environment of the Baltic Sea area from all sources of pollution, and to restore and safeguard the ecological balance of the Baltic Sea area. The revised 1992 Convention places greater focus on nature conservation and the protection of biological diversity including conserving natural habitats and protecting ecological processes to ensure sustainable use of natural resources
1979	Bonn Convention on the Conservation of Migratory Species of Wild Animals, including 1991 ASCOBANS to protect and conserve small cetaceans in Baltic Sea and North Sea
1979	Bern Convention for Conservation of European Wildlife (fauna and flora) and Natural Habitats, giving particular emphasis to endangered or vulnerable species including those that are migratory species
1982	UN Convention on the Law of the Sea (UNCLOS) identifying rights and responsibilities of States regarding resource management and protection of the marine environment
1992	Convention on Biological Diversity (CBD) promoting conservation of biological diversity, sustainable use of its components, and fair and equitable sharing of the benefits from the use of genetic resources
1992	UN Framework Convention on Climate Change (UNFCCC) sets an overall framework for intergovernmental efforts to tackle the climate change challenge. It recognizes that the climate system is a shared resource whose stability can be affected by industrial and other emissions of heat-trapping 'greenhouse' gases. The Kyoto Protocol, an international, legally binding agreement to reduce worldwide the emissions of greenhouse gases, came into force on 16 February 2005

(continued)

Table 1.1 (continued)

Year	Instrument
1995	FAO Code of Conduct on Responsible Fisheries providing guidelines and principles promoting responsible fishing and fisheries activities, taking into account relevant biological, technological, economic, social, environmental and commercial aspects
1995	UN Convention on Straddling Fish Stocks and Highly Migratory Fish Stocks seeks to lay down a comprehensive regime for the conservation and management of such stocks
1998	UN Economic Commission for Europe (UNECE) Aarhus Convention ensuring access to information, public participation in decision-making and access to justice in environmental matters
2004	International Maritime Organization (IMO) International Convention for the Control and Management of Ships' Ballast Water and Sediments to prevent the spread of harmful aquatic organisms
2009 – ongoing	Agreement between the European Community/European Union and the Government of the Russian Federation on cooperation in fisheries and the conservation of the living marine resources in the Baltic Sea. Aims to ensure that the conservation, exploitation and management of any straddling, associated and dependent stocks provides sustainable, economic, environmental and social conditions

The list is illustrative rather than definitive

1.3.2 The Primary EU Environmental Instruments in Coastal and Offshore Environments

For the EU countries, including those around the Baltic Sea, the main instruments of marine environmental policy are:

- The Water Framework Directive.
- The Marine Strategy Framework Directive.
- The Common Fisheries Policy.
- The Habitats and Birds Directives.

These are examined in more detail below.

1.3.2.1 The Water Framework Directive

The Water Framework Directive (WFD) established a new legal framework for the protection, improvement and sustainable use of surface waters, transitional waters, coastal waters and groundwater across Europe, so as to prevent deterioration, enhance status of aquatic ecosystems, including groundwater and reduce pollution (EC 2000). The directive repeals and replaces a number of older directives and incorporates the remaining existing water directives (Bathing Water, Nitrates and Urban Waste Water Treatment Directives) into its framework through its protected areas provisions. The WFD requires, inter alia, the establishment of classification schemes to reflect the

Table 1.2 Some European Economic Community (EEC)/European Community (EC) instruments applicable to the protection of the environment and biodiversity, including conservation of fishery resources, of the Baltic Sea region

Year	Instrument
1976	Directive on Bathing Water (76/160/EEC) sets cleanliness standards for bathing water
1976	Directive on Dangerous Substances (76/464/EEC) aims to reduce or eliminate pollution from chemicals
1979	Directive on the Conservation of Wild Birds (79/409/EEC) establishes special protection areas and conservation measures to protect habitats of rare or vulnerable species including migrants
1979	Directive on Shellfish Growing Waters (79/923/EEC) aims to protect and, where necessary, improve the quality of waters where shellfish grow and contribute to high quality of edible shellfish products
1970(83) – ongoing	Common Fisheries Policy (CFP) with first common measures dating from 1970, applying to both fisheries and aquaculture, was adopted in 1983 by introducing a fisheries conservation policy to complement the already adopted structural, market and external policies. It aims to ensure exploitation of living aquatic resources providing sustainable economic, environmental and social conditions. Recent CFP reforms have inter alia aimed at greater integration of environmental and biodiversity concerns into the CFP and implementation of the ecosystem approach to management (EAM)
1985/97	Environmental Impact Directive (85/337/EEC superseded by 97/11/EC) requires developer to provide information to competent authority about likely significant environmental effects
1991	Directive on Aquaculture Animals and Products (91/67/EEC) aims to increase productivity, introduce health rules, and limit the spread of infections and contagious diseases
1991	Urban Waste Water Treatment Directive (91/271/EEC) addresses the major point sources, namely the municipal waste water discharges, and sets minimum standards for the collection, treatment, and disposal of waste water dependent on the size of the discharge
1991	Nitrates Directive (91/676/EEC) concerns the protection of surface and ground waters from pollution from the application/storage of inorganic fertilizers and manure from diffuse agricultural sources
1992	EC Directive on Conservation of Natural Habitats and Wild Fauna and Flora (92/43/EEC) is the means by which the EC meets its obligations as a signatory of the Bern Convention. Designates and implements conservation measures for Special Areas of Conservation
1997	Amsterdam Treaty, strengthens previous environmentally orientated declarations, sets environmental policy objectives by explicitly stating, as EC tasks, 'a high level of protection and improvement of the quality of the environment'
2000	Water Framework Directive (2000/60/EC) (WFD) promotes the integrated management of all water-related operations in fresh and marine waters, including coastal waters. The WFD applies the EAM, and enhances conservation and where necessary recovery and restoration of ecosystems and their habitats, by establishing measures to terminate or phase out discharges, emissions and losses of pollutants including nutrients, with the ultimate aim of achieving concentrations in the marine environment near background values for naturally occurring substances and close to zero for man-made synthetic substances. The WFD requires classification of the quality status of coastal marine waters. By 2013, several aspects of EC water legislation will be streamlined and subsumed by the WFD

(continued)

Table 1.2 (continued)

Year	Instrument
2006 ongoing	The Marine Strategy Framework Directive (MSFD) 'Establishing a Framework for Community Action in the field of Marine Environment Policy' (2008/56/EC) aims to cover the actions needed to ensure that all human activities impacting the oceans and seas are managed so that marine biological diversity and critical habitats are conserved and human use of them is sustainable. The MSFD applies an integrated EAM to achieve good environmental status of marine waters by 2021 and to protect the resources upon which marine-related economic and social activities depend. The MSFD is the environmental pillar of the Maritime Policy (cf. Green Paper for an EU Maritime Policy (COM (2006) 275) and Blue Book 'An Integrated Maritime Policy for the European Union' (COM (2007) 575 final)), designed to achieve the full economic potential of oceans and seas in harmony with the marine environment. Each Member State, cooperating with relevant other Member States and third countries within a region, must develop Marine Strategies for its marine waters. The strategies require a detailed assessment of environmental status, a definition of 'good environmental status', and establish clear environmental targets and monitoring programmes. States must draw up programmes of cost-effective and impact assessed measures

The list is illustrative rather than definitive

ecological status or potential of surface water bodies and coastal waters as measured by the condition of specific biological, hydromorphological and physico-chemical elements. Within the WFD, the coastal waters are first typified and then classified by the Member States into five ecological quality status classes (i.e. high, good, moderate, poor and bad) with the ultimate objective of achieving good status for surface waters. A set of guidelines and methodology for classification of coastal waters has been developed. In the inter-calibration process, Member States decide on interpretation of the environmental objectives.

The basic aim of achieving 'good ecological status' of the WFD is very similar to those in the MSFD although there are some key differences:

1. The WFD is already in its implementation phase so the timetable is much tighter. Monitoring has started and the first River Basin Management Plan should be completed by 2009, thereby defining the environmental objectives and programme of remedial measures. A further plan has to be produced by 2013 with the aim of achieving good surface water status by 2015. The cycle is then repeated over 6-year periods;
2. The spatial scale of the WFD only covers inshore coastal waters and transitional waters (i.e. brackish waters and estuaries) out to 3 nm from the shore-related baseline;
3. Importantly, the WFD brings in holistic regulations and monitoring programmes that cover the land/sea interface. As most pollution entering the sea emanates from land-based sources, control at this point is most effective.

An important part of WFD implementation is international inter-calibration work to ensure that 'good ecological status' is equivalently defined and implemented by all Member States. Similar work is needed under the MSFD.

1.3.2.2 The Marine Strategy Framework Directive

The Marine Strategy Framework Directive (MSFD) (EC 2008a) will effectively extend many of the provisions of the WFD to marine areas/seas beyond the current WFD limits. The MSFD requires (the author has applied bold text to emphasize particularly important aspects) Member States to *take the* ***necessary measures to achieve or maintain good environmental status in the marine environment by the year 2020 at the latest.***

To deliver this, EU Member States are required to develop and implement Marine Strategies (i.e. Action Plans) in order to:

1. ***Protect and preserve the marine environment, prevent its deterioration or, where practicable, restore marine ecosystems in areas where they have been adversely affected***; [Art1(2)(a)]
2. ***Prevent and reduce inputs in the marine environment, with a view to phasing out pollution*** *as defined in Article 3(8), so as* ***to ensure that there are no significant impacts on or risks to marine biodiversity, marine ecosystems, human health or legitimate uses of the sea***; [Art1(2)(b)]
3. *Marine Strategies shall* ***apply an ecosystem-based approach to the management of human activities, ensuring that the collective pressure of such activities is kept within levels compatible with the achievement of good environmental status and that the capacity of marine ecosystems to respond to human-induced changes is not compromised, while enabling the sustainable use of marine goods and services by present and future generations***. [Art1(3)]

The EU Member States must undertake an initial assessment of their waters consisting of: (a) an analysis of their essential physical and biological features and the environmental status of marine waters; (b) an analysis of the predominant pressures and impacts that might adversely affect the environment; and (c) a socio-economic analysis of their use and the costs of degradation, i.e. a form of cost – benefit analysis. The list of characteristics should cover the habitat types, biological components, physico-chemical components and other features. The approach is consistent with the selection of hydro-morphological and biological elements under the WFD, although the focus on maintenance of water quality in the WFD requires monitoring of only macrophytes, macroalgae, phytoplankton, macrobenthos and fish in transitional waters. Less comprehensive approaches are also taken by the Habitats and Birds Directives and Annex V of the OSPAR Convention, which provide lists of species and habitats as specific items for development of management measures and as proposals for Ecological Quality Objectives (EcoQOs) in the OSPAR context.

The MSFD's schedule of major milestones (Fig. 1.1) consists of: (a) an *initial assessment* of the current environmental status of marine waters (by July 2012); (b) a determination of *what constitutes Good Environmental Status* (GES) for those waters, including establishment of environmental targets and indicators for achievement or maintenance of GES (by July 2012); (c) establishment and

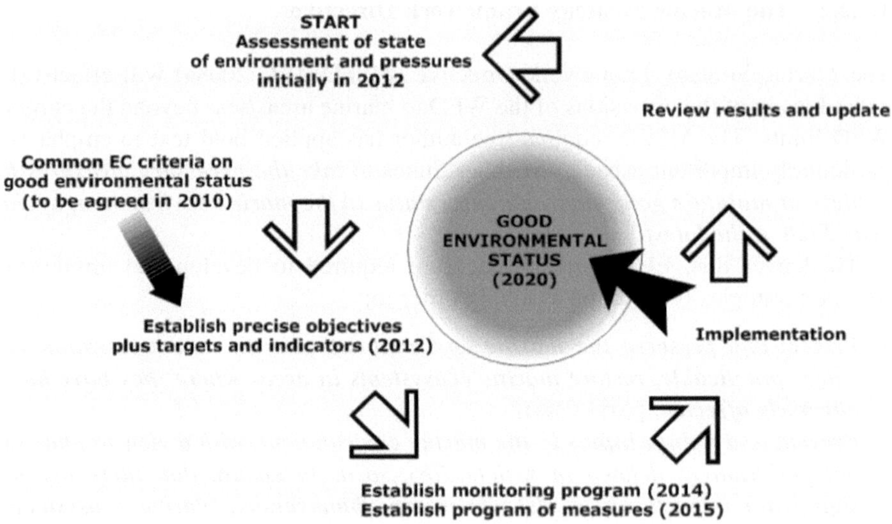

Fig. 1.1 Schedule for implementing the main milestones of the Marine Strategy Framework Directive

implementation of a *monitoring programme* to measure the current status and progress towards achieving GES (by July 2014); and (d) establishment of a *programme of measures* (e.g. management actions concerning human activities) (by 2015) for both land-based and seas-based pressures, designed to achieve or maintain GES (by 2020 at the latest).

For the MSFD, GES means: 'the environmental status of marine waters where these provide ecologically diverse and dynamic oceans and seas which are clean, healthy and productive within their intrinsic conditions, and the use of the marine environment is at a level that is sustainable, thus safeguarding the potential for uses and activities by current and future generations', i.e.:

(a) 'The structure, functions and processes of the constituent marine ecosystems, together with the associated physiographic, geographic, geological and climatic factors, allow those ecosystems to function fully and to maintain their resilience to human-induced environmental change. Marine species and habitats are protected, human-induced decline of biodiversity is prevented and diverse biological components function in balance.
(b) Hydro-morphological, physical and chemical properties of the ecosystems, including those properties which result from human activities in the area concerned, support the ecosystems as described above. Anthropogenic inputs of substances and energy, including noise, into the marine environment do not cause pollution effects.'

The preceding text for GES is basically comparable with the WFD's definition of good ecological status (GEcS) for coastal waters.

The MSFD provides 11 high-level, qualitative descriptors for 'GES' to be achieved:

1. Biological diversity is maintained. The quality and occurrence of habitats and the distribution and abundance of species are in line with prevailing physiographic, geographic and climatic conditions;
2. Non-indigenous species introduced by human activities are at levels that do not adversely alter the ecosystems;
3. Populations of all commercially exploited fish and shellfish are within safe biological limits, exhibiting a population age and size distribution that is indicative of a healthy stock;
4. All elements of the marine food webs, to the extent that they are known, occur at normal abundance and diversity and levels capable of ensuring the long-term abundance of the species and the retention of their full reproductive capacity;
5. Human-induced eutrophication is minimized, especially adverse effects thereof, such as losses in biodiversity, ecosystem degradation, harmful algal blooms and oxygen deficiency in bottom waters;
6. Sea floor integrity is at a level that ensures that the structure and functions of the ecosystems are safeguarded and benthic ecosystems, in particular, are not adversely affected;
7. Permanent alteration of hydrographical conditions does not adversely affect marine ecosystems;
8. Concentrations of contaminants are at levels not giving rise to pollution effects;
9. Contaminants in fish and other seafood for human consumption do not exceed levels established by Community legislation or other relevant standards;
10. Properties and quantities of marine litter do not cause harm to the coastal and marine environment;
11. Introduction of energy, including underwater noise, is at levels that do not adversely affect the marine environment.

The importance of Marine Protected Areas (MPAs) in achieving GES is also recognized.

The EU Member States must determine how the MSFD will be implemented, via allocating responsibilities to competent national agencies, and in close cooperation with other Member States and third parties (e.g. Russia) in appropriate regional forums (e.g. HELCOM, OSPAR). Regional assessments – covering so-called European Marine Regions (see Sect. 1.3.3), such as the Baltic Sea – will be delivered through the auspices of the regional forums.

EU standardized (e.g. common and comparable) monitoring and assessment methods will be developed and applied; wherever possible using or building on approved methodology.

The MSFD stops short of proposing what specific measures Member States should bring forward to achieve GES. However, the MSFD expects that Member States sharing a marine region shall cooperate to ensure that the measures required to achieve the objectives of the MSFD are coherent and coordinated across the marine region. The European Commission has produced a 'Decision' providing guidance on the development of common criteria and methodological standards for defining GES of marine waters (EC 2010) in the context of Annex 1 of the MSFD. The 'Decision' inter alia emphasizes the need for a range of focused research and technological development (RTD) – in the form of knowledge (basic and applied) and data collection – about various species, populations, habitats and ecosystems in order to achieve the monitoring and assessment requirements which lie at the heart of the MSFD.

1.3.2.3 The Habitats and Birds Directives

The Habitats (EEC 1992) and Birds (EEC 1979) Directives require EU Member States to identify and designate two types of area: (1) Special Areas of Conservation (SAC), where they support rare, endangered or vulnerable natural habitats and species of plants and animals (other than birds) as listed in Annexes 1 and 2 of the Habitats Directive; and (2) Special Protection Areas (SPA) that support significant numbers of wild birds and their habitats as listed in the Birds Directive. The requirements of these directives extend out to the 200 nm limit. The collection of a representative, ecological coherent network of well-managed sites, known as Natura 2000, is designed not only to ensure the survival of important habitats and species but also encourages the sustainable management of the marine environment as a whole.

It is especially noteworthy that a specific requirement of the MSFD is that marine strategies must include the creation by Member States of 'spatial protection measures, contributing to coherent and representative networks of marine protected areas', such as those required under the Birds and Habitats Directives and other (e.g. non-EU) international obligations, by 2013.

1.3.2.4 The Common Fisheries Policy

The origin and development of the Common Fisheries Policy (CFP) up to 1994 is described by Holden (1994). The 2002 reform and ongoing periodic reform, due by 2012, of the CFP have emphasized that the EAM must be fully integrated and implemented into the principles, objectives and operational framework of the CFP (Sissenwine and Symes 2007; EC 2009a) and the new overarching Integrated Maritime Policy (EC 2007), under which research in support of policy, scientific evidence-based advice and management regarding capture fisheries play essential roles. There is an emphasis shift from the current primarily *tactical*, year-to-year (i.e. short-term) single stock management approach towards a longer term *strategic* approach. This involves credible strategies for establishing multi-year management plans including effective recovery plans for depleted stocks, which take better account of multispecies interactions in the ecosystem, relevant human dimensions (e.g. socio-economics, governance), tackling by-catch/discarding and illegal catches, mitigation/adaptation to climate change, and a priori evaluation of the efficacy of proposed strategies (EC 2009a). A fundamental challenge is to avoid overexploitation by achieving an appropriate balance between fishing effort/mortality and the available living resources, thereby increasing the potential for achieving long-term maximum sustainable yields and also substantially limiting the ecosystem impacts.

Whereas substantial progress has been made to reduce and prevent many kinds of pollution of the marine environment in the European regional seas (Nilsen et al. 2002; ICES 2003; Hopkins 2004; GIWA 2005), the European Commission has acknowledged that the CFP has not been successful as measured by 88% of the fish stocks being overfished and 30% outside 'safe biological limits' (EC 2009a).

There is a need to better integrate the CFP with other marine governance regimes such as environmental policy (e.g. MSFD) and with other sectors as outlined in the EU Integrated Maritime Policy (EC 2006, 2009a). In particular, the MSFD (EC 2008a) states:

> 'Where a Member State identifies an issue which has an impact on the environmental status of its marine waters and which cannot be tackled by measures adopted at national level, or which is linked to another Community policy [e.g. CFP] or international agreement, it shall inform the Commission accordingly and provide a justification to substantiate its view [Art15(1)]'. The Commission is required to respond within 6 months [Art13(1) & 14(1)].

Thus, Member States will still rely on the adoption of an EU Regulation to prohibit fisheries where they are a damaging activity [Recital 39], but there is now an implicit acknowledgement of the need for the Commission to act and respond within a relatively short period of time. Recital 40 requires that the CFP and any future reform of it, 'should take into account the environmental impacts of fishing and the objectives of this Directive'.

1.3.3 Marine Regions: A Basis for Environmental Management

Ecosystem-based management is place- or area-based in focusing on a specific ecosystem and cohesively managing the range of human activities affecting it (JMM 2003). Marine 'ecoregions', also known as Large Marine Ecosystems (LMEs) are regions of the world's seas and oceans identified to aid in conservation activities for marine ecosystems (Sherman 1994; Sherman and Duda 1999; Dinter 2001; Suárez de Vivero et al. 2009). Thus, the identification of an ecoregion may take account of inter alia biogeographic and oceanographic features, with due regard to links between the marine and terrestrial environment, including patterns of land use and distribution and density of human populations, as well as to existing political, social, economic and management divisions. In essence, marine ecoregions are nested within a biogeographically based framework: they encompass distinctive biological communities, having characteristic ecological processes and productivity, which are under the influence of distinguishing physical and chemical oceanographic regimes (cf. Sherman 1994; Longhurst 1998; Hopkins 2004). The fluid nature of water and ocean currents, and the dispersal patterns of many plants and animals, results in the patterns of biodiversity being 'transboundary' and not conforming to national waters and national exclusive economic zones (EEZs). Thus, conservation at an ecoregional scale and other large-scale approaches are essential for the successful management of the marine environment including its biodiversity. Accordingly, ecoregional conservation provides a framework for applying the 'ecosystem approach to management' at a variety of scales.

The Baltic Sea has been designated as LME No. 23 (Heileman and Thulin 2009) of a total of 64 LMEs worldwide. Initial science-based management of the Baltic Sea LME was first established in 2003 under the auspices of ICES (Thulin 2009). The LME concept for monitoring, assessment and management of international coastal waters was conceived in the 1980s, and has been developed and further refined as a complementary instrument world-wide for achieving an EAM (Sherman 1994; Sherman and Duda 1999; NOAA 2005).

The MSFD will be applied in the European regional seas at the level of four specific 'European Marine Regions' and their respective sub-divisions (Suárez de Vivero et al. 2009): (1) North East Atlantic, comprising four sub-regions, totaling 4,673,125 km^2; (2) Mediterranean Sea, comprising three sub-regions, totaling 1,533,098 km^2; (3) Baltic Sea, with no currently specified sub-regions, totaling 349,644 km^2; and (4) Black Sea, with no currently specified sub-regions, totaling 55,908 km^2.

In operational terms, the MSFD appears to recognize 'ecosystem-based management units' which form a compromise between 'ecoregions' and 'management regions'.

1.3.4 Baltic Sea Marine Environmental Management: HELCOM, EU and Russia

In the Baltic Sea, commercial fisheries management between the EU and Russia occurs on the basis of a bilateral agreement. In the case of 'environmental' management excluding commercial fisheries management, all Baltic Sea countries collaborate primarily via HELCOM.

1.3.4.1 Fisheries: The EU – Russia

The agreement between the EC/EU and Russia on cooperation in fisheries and the conservation of living marine resources in the Baltic Sea came into force on 29 April 2009 (EC 2009b). The agreement, which covers an initial period of 6 years, sets out provisions on joint management measures for Baltic Sea fisheries. The bilateral framework became necessary as, following the 2004 EU enlargement, the membership of the International Baltic Sea Fisheries Commission (IBSFC, the Convention on Fishing and Conservation of the Living Resources in the Baltic Sea and the Belts of 1973, the so-called Gdansk Convention) consisted of only two Parties, i.e. the EU Member States and Russia. The IBSFC ceased to exist on 1 January 2007, following the withdrawal of the EU from the Gdansk Convention.

The Baltic Sea Regional Advisory Council (BS RAC) was set up in March 2006 in accord with Council Decision 2004/585/EC of July 2004 to create so-called RACs in support of the 2002 reform of the CFP. The RACs originated as a response to calls from stakeholders in the fisheries sector who wanted to be more involved in the way fisheries are managed in the EU. The main aim of the BS RAC is to prepare and provide advice to the European Commission and Member States on matters relating to the management of Baltic Sea fisheries in order to achieve a successful running of the EU's Common Fisheries Policy. The BS RAC consists of representatives from the fishing sector and other interest groups affected by the CFP. The BS RAC is one of seven RACs established by the European Council to increase stakeholder involvement in the development and governance of a successful CFP.

1.3.4.2 'Environment': HELCOM and the BSAP – Forum for All Baltic Sea Countries

The 1992 Helsinki Convention (replacing the 1974 convention), supported by its governing body the Helsinki Commission (HELCOM), has as its main goal the protection of the marine environment of the Baltic Sea from all sources of pollution, and to restore and safeguard its ecological balance (HELCOM 1992; Ehlers 1994). The mission of HELCOM requires extensive implementation of the EAM, the precautionary principle and related concepts. The current Contracting Parties include all nine Baltic coastal countries as well as the EU. HELCOM, besides the Heads of Delegation, comprises five operational groups: Response (HELCOM RESPONSE); Monitoring and Assessment (HELCOM MONAS); Maritime (MARITIME); Land-based Pollution Group (LAND); and Nature Protection and Biodiversity (HABITAT). The EAM adopted by the 2003 Joint HELCOM/OSPAR Ministerial Meeting (JMM 2003), obliges HELCOM to assess the pressures as well as the resulting state and impacts on the marine environment and to use this as the foundation for priority actions. The new HELCOM Monitoring and Assessment Strategy, adopted in 2005, aims inter alia to: (a) facilitate the implementation of the EAM; (b) show the interlinkage and interdependence of activities on land, in coastal areas and at sea; (c) coordinate monitoring activities for Baltic specific issues of concern; and (d) produce targeted environmental assessments for regional specific management purposes by also making use of data and information produced by Contracting States and other and other forums.[3] The HELCOM monitoring programmes cover eutrophication, hazardous substances and elements of

[3] *Inter alia*: EEC/EC directives and other instruments (e.g. Water Framework Directive, Urban Waste Water Treatment Directive, Nitrates Directive, Habitat Directive, Dangerous Substances Directive, Integrated Pollution Prevention Control (IPPC) Directive, European Marine Strategy), European Environment Agency, International Atomic Energy Authority, ICES, Organisation for Economic Cooperation and Development, and UN Environment Programme.

biodiversity. Baltic Sea Protected Areas (BSPAs) provide building blocks for nature conservation assessments. Since 1981, HELCOM has conducted periodic assessments of the status of the Baltic marine environment.

The Baltic Sea Action Plan (BSAP), adopted at a 2007 HELCOM Ministerial Meeting, is an ambitious programme to address all the major environmental problems affecting the Baltic marine environment, and restore the good ecological status of the Baltic marine environment by 2021 (HELCOM 2007b). The BSAP aims to safeguard the region's natural ecosystems while also allowing valuable marine resources to be used sustainably in the future. The social and economic effects of the measures proposed for the plan will be carefully assessed, as well as their beneficial impacts on the environment. The BSAP implements obligations under other international legislative frameworks, such as the CBD, the 2002 WSSD, and the EU Maritime Policy with its supporting MSFD, and WFD. The BSAP forms an important contribution to the MSFD, which foresees a separate action plan for each marine region. The four main objectives of the BSAP are to achieve: (1) a Baltic Sea unaffected by eutrophication; (2) a Baltic Sea undisturbed by hazardous substances; (3) a Baltic Sea with environmentally friendly maritime activities; and (4) a favourable conservation status of Baltic Sea biodiversity. The BSAP not only substantially increases the prospects for future recovery of the Baltic Sea marine environment but also improves the environment's capability to react to the projected stress resulting from climate change.

The nine Baltic Sea countries have had to establish, by 2010, national programmes in support of the BSAP and their effectiveness is to be evaluated at a Ministerial Meeting in 2013.

More detailed information about HELCOM and the BSAP is provided in this book by Minna Pyhälä (Chap. 2).

1.3.4.3 ICES: Providing Independent Scientific Advice for Management

The International Council for the Exploration of the Sea (ICES), founded as an intergovernmental organization in 1902, coordinates and promotes marine research on oceanography, the marine environment, the marine ecosystem, and on living marine resources in the North Atlantic (Rozwadowski 2002). Members of the ICES community now include all coastal states bordering the North Atlantic and the Baltic Sea, with affiliate members in the Mediterranean Sea and southern hemisphere. ICES is a network of more than 1,600 scientists from 200 institutes linked by an intergovernmental convention (ICES 1964) to add value to national research efforts.

ICES is the prime source of independent, objective scientific advice on the marine ecosystem to governments and international regulatory bodies that manage the North Atlantic and adjacent seas. In the Baltic Sea, ICES provides scientific advice to all coastal countries, HELCOM and the European Commission.

1.4 The Key Policy and Management Aim: Conserving Healthy and Sustainable Ecosystems

The key marine environmental policy and management aim is to conserve healthy and sustainable ecosystems. In order to achieve this aim the main instrument is the development and application of the EAM, which is fortified by the precautionary principle and the polluter pays principle.

1.4.1 The Precautionary Principle and the Polluter-Pays Principle

Many international agreements promote the requirement for the prudent and sustainable use of the marine environment, biodiversity including living marine resources, and ecosystems (cf. Tables 1.1, 1.2). These recognize that in order to achieve sustainability it is necessary to balance the effects of exploitation with the needs of conservation. Thus, the *precautionary principle* and the *polluter-pays principle* are promoted as key principles underpinning the EAM.

The precautionary principle (PP) – and the way to implement it (the precautionary approach) – have arisen from the Rio Declaration (UNCED 1992), which states that 'where there are threats of serious or irreversible damage, lack of full scientific certainty shall not be used as a reason for postponing cost-effective measures to prevent environmental degradation'.

The PP recognizes that changes in exploited systems are slowly reversible, difficult to control, not well understood, and calls for early action in the case of uncertainty and ignorance (i.e. incomplete knowledge) in order to prevent potential harm. Regarding precaution, scientific advice and management should have safe margins of error and avoid risk, and appropriate management plans and recovery plans need to be developed and implemented in this context. In particular, besides developing a timely Code of Conduct for Responsible Fisheries (FAO 1995), FAO (1996) developed supplementary guidelines for application of the precautionary approach.

Precaution places the *burden of proof* on the proponents of the activity, i.e. it is not appropriate to assume that environmental impacts are negligible until proved otherwise. This reversal of the burden of proof is fundamental to precautionary action, and creates incentives for the proponents of an activity to prove that their product or activity is safe or acceptable. In the EU, this may be achieved, for example, through application of an environmental impact assessment (EIA) and strategic environmental impact assessment (SEA).

The *polluter-pays principle* (PPP) asserts that the costs of environmental damage or resource depletion should be borne by polluters or users. It was elaborated as an economic principle in the 1970s and is embedded in the 1987 European Community (EC) Treaty known as the Single European Act (EC 1987). The PPP requires producers or resource users to meet the cost of implementing environmental standards or technical regulations, or by introducing liability regimes making producers liable for causing environmental damage (Sadeleer 2007). The PPP establishes a framework based on environmental liability covering damage to the biodiversity protected at EU and national levels, to waters regulated by the WFD (EC 2000) and the MSFD (EC 2008a), as well as air and land contamination which cause serious harm to human and ecosystem health.

The North Sea Conference (NSC) process has become a world-wide trend-setter by introducing the principle of precautionary action. It has led to the ban on dumping and incineration of industrial waste at sea, the cessation of sewage sludge dumping, the land-based disposal of disused offshore installations and the reduction of inputs of certain pollutants from land-based sources. These commitments by the North Sea Environment Ministers have subsequently become central in legislation connected with the European regional seas, such as the 1992 HELCOM Convention concerning the Baltic Sea and the 1992 OSPAR Convention on the Protection of the Marine Environment of the North–East Atlantic, with 1998 Annex V on 'The Protection and Conservation of the Ecosystems and Biological Diversity of the Maritime Area'.

In adopting the target of the cessation of discharges, emissions and losses of hazardous substances within one generation (*ca.* 25 years), the Ministerial Declaration from the Fourth International Conference on the Protection of the North Sea (4NSC) (NSC 1995) delivered a visionary milestone in terms of aquatic and chemicals policy which now prevails throughout the EU:

> The Ministers agree that the objective is to ensure a sustainable, sound and healthy North Sea ecosystem. The guiding principle for achieving this objective is the precautionary principle. This implies the prevention of the pollution of the North Sea by continuously reducing discharges, emissions and losses of hazardous substances thereby moving towards the target of their cessation within one generation (25 years) with the ultimate aim of concentrations in the environment near background values for naturally occurring substances and close to zero concentrations for man-made·synthetic substances.

1.4.2 The Ecosystem Approach to Management and the 2002 Bergen Ministerial Declaration

1.4.2.1 The Ecosystem Approach: Paradigm Shift Towards Sustainability of Ecosystem Goods and Services

The *EAM* – also synonymous with 'ecosystem approach', 'ecosystem-based management', etc. – has been adopted as the fundamental tool for attaining sustainable development and has been integrated into the conventions, agreements and work-plans of the implementing international commissions and international organizations. Importantly, the EAM is the primary framework for action under the CBD (1992), in which the 12 'Malawi Principles' (CBD 1998) emphasize that humans are integral components of ecosystems whereby human social and economic systems constantly interact with the physical, chemical and biological parts of the ecosystem. Thus, the EAM is concerned with managing human impacts on ecosystems rather than implying an attempt to manage ecosystems by humans, in order to seek an appropriate balance between conservation and use of biological diversity. Consequently, the ecosystem approach is a holistic process for integrating and delivering in a balanced way the three objectives of the CBD regarding biodiversity: (1) conservation; (2) sustainable use; and (3) equitable sharing of the benefits. The ecosystem approach requires adaptive and integrated management of land, water, and living resources to deal with the complex and dynamic nature of ecosystems and the absence of complete knowledge or understanding of their functioning. Thus, the EAM requires a comprehensive and holistic approach to understanding and anticipating ecological change, assessing the full range of consequences and developing appropriate scientific advice and knowledge-based management and regulatory responses. Implementing the EAM is an adaptive and evolutionary process and should be considered as a *tool* to help comprehensively and systemically *redress the causes of human induced problems* affecting the sustainability of ecosystems (Hopkins 2004; Rice et al. 2005).

Healthy ecosystems perform a diverse array of essential functions that provide both goods and services to humanity. ‘Goods’ refers to items given monetary value in the market place and ‘services’ from ecosystems are valued but rarely bought and sold (Lubchenco 1994; Nilsen et al. 2002):

- ‘Goods’, for example, are: food (fisheries and aquaculture); medicinal materials; raw materials (oil and gas) and wild genes;
- ‘Services’, for example, are: maintaining the hydrological cycles and composition of the atmosphere; regulating climate; storing and cycling essential nutrients; absorbing and detoxifying pollutants; providing sites for tourism, recreation, education and research.

The sustainability concept depends on two aspects: (a) sustainability of use (sustainable use); and (b) sustainability of ecological resources and their associated ecosystem. These aspects are tightly connected because sustainable use of ecological resources can only be achieved if these resources themselves are sustainable. Thus, the EAM involves *a paradigm shift* from managing commodities associated with traditional sectoral industries towards sustaining the production potential for both ecosystem goods and services (‘natural capital’), in which most of the services have consistently been overlooked and undervalued (Costanza et al. 1997).

1.4.2.2 The 2002 Ministerial Conference on the North Sea and Its Aftermath in the European Regional Seas

The Fifth International Conference on the Protection of the North Sea (5NSC) led the way to recognizing the need to manage all human activities that affect the North Sea, and indeed other European regional seas, in an integrated way that conserves biological diversity, and applies the EAM to ensure sustainable development. As part of the 5NSC process, the 1997 Intermediate Ministerial Meeting on the

Integration of Fisheries and Environmental Issues in the North Sea (IMM 1997) recognized the desirability of an *ecosystem approach* (Nilsen et al. 2002), and the 2002 Ministerial Declaration from 5NSC, the so-called Bergen Declaration (NSC 2002), agreed 'to implement an ecosystem approach by identifying and taking action on influences which are critical to the health of the North Sea ecosystem, and that management will be guided by a conceptual framework which includes:

- The development of general and operational environmental goals;
- Best use of available scientific knowledge about the structure and function of the ecosystem;
- Best use of scientific advice;
- Integrated expert assessment;
- Coordinated and integrated monitoring;
- Involvement of all stakeholders;
- Policy decisions and control and enforcement.'

To implement an EAM in line with this framework the following aspects were emphasized as important for success (NSC 2002):

1. Develop focused research and information gathering which addresses the driving forces of ecosystems variability, including climatic, biological and human factors which are critical for maintaining ecosystem structure and function;
2. The need for shared integrated expert advice and assessments, including marine resources, environmental and socio-economic factors;
3. Develop a strategy for achieving dialogue with all relevant stakeholders for the development and implementation of the ecosystem approach;
4. Improve the coordination, harmonization and efficiency of current national and international monitoring to serve the assessment processes;
5. Make appropriate policy decisions, including integration of environmental protection into all sectors, implement the corresponding management actions, and ensure proper control and enforcement to deliver an ecosystem approach; and
6. Make use of ecological quality objectives (EcoQOs) as a tool for setting clear, operational environmental objectives directed towards specific management and serving as indicators for the ecosystem health.

Following on from 5NSC, the EAM was officially defined and agreed, at the First Joint Meeting of the HELCOM and OSPAR Commissions in June 2003 (JMM 2003) as 'the comprehensive integrated management of human activities based on the best available scientific knowledge about the ecosystem and its dynamics, in order to identify and take action on influences which are critical to the health of marine ecosystems, thereby achieving sustainable use of ecosystem goods and services and maintenance of ecosystem integrity'. This definition of the EAM has been officially adopted throughout the European Community and its subsequent replacement in the form of the EU in, for example, the MSFD.

Developing and implementing the EAM, with elaboration of suites of EcoQOs, has since become central to the protection of the North Sea, and the work of OSPAR and

HELCOM (NSC 2002; JMM 2003) in the North–East Atlantic and the Baltic Sea, respectively.

For the European regional seas, EcoQOs are being developed to provide objectives, and thus operational tools, for implementing the EAM. The main aspects of the EAM framework, including application of EcoQOs, are examined below. It must be emphasized that the EAM is an evolutionary and adaptive process representing the primary forum for linking science and policy (Rice et al. 2005; Hopkins et al. 2006) (see the following section and Sect. 1.6).

1.4.3 The EAM Framework

The EAM framework needs a vision for conservation of, or attaining should it be in a degraded state, a 'healthy' ecosystem, setting ecological objectives related to this, monitoring and assessing indicators of change, and measures connected with the management/regulation of the human pressures causing undesired impacts on the ecosystem.

1.4.3.1 What Has 'Health' Got to Do with It?

Ecosystem 'health' (well-being) has several analogies with human health related to the body's components (e.g. organs). The ecosystem is made up of components whose individual 'health' should be maintained in an integrated manner at various desirable levels of 'quality' to ensure that the whole ecosystem is healthy and functions well regarding the provision of 'goods and services'. Thus consider:

- Health: 'a state in which you are fit and well' (BBC 1993);
- Regarding an organization/system (cf. ecosystem): 'working well' (i.e. sum of the parts);
- Healthy: The extent a person/something is in 'good condition' as a whole (some organs/systems may be healthy and others not). Good health and its deviation are measured against various 'reference levels' (e.g. normal body temperature and blood pressure);
- When condition deviates *significantly* from *desired target level of good health* it is implicit that one has crossed a limit level beyond which serious/irreversible harm results;
- To avoid serious/irreversible harm, one should take precautions and establish *precautionary levels* beyond which we should not transgress (cf. 'body mass index' as a simple precautionary indicator for obesity). Keeping within the precautionary levels is wise: it avoids major socioeconomic costs that arise from ill-health and its treatment (cf. recovery).

1.4.3.2 The Need for a Vision: A Benign Ambition of What One Wishes to Achieve

A convincing and coherent ecosystem-related policy (e.g. the MSFD) incorporates a clear overarching *vision*, as manifested in the following aspirations: (a) *MSFD*: 'maintaining biodiversity and providing diverse and dynamic oceans and seas which are clean, healthy and productive.' (EC 2008a); (b) *OSPAR*: 'a clean, healthy, biologically diverse North–East Atlantic ecosystem' (OSPAR 2010); and (c) *HELCOM*: 'a healthy marine environment, with diverse biological components functioning in balance, resulting in a good ecological status and supporting a wide range of sustainable human activities' (HELCOM 2007b).

1.4.3.3 Human Pressures, Ecological Issues and Ecological Quality Objectives: Where Are We, Where Have We Come from, and Where Do We Wish to Go?

Human Activities and Pressures

Particularly since the industrial revolution, and manifestly today, diverse human activities have not only benefited from but also increasingly exerted pressures on marine ecosystems, including the Baltic Sea ecosystem (ICES 2003; Hopkins 2004; GIWA 2005; Hopkins et al. 2006) (Fig. 1.2).

The human-induced pressures have contributed to:

- Intensive and unsustainable exploitation of many fish stocks and other living resources;
- Pollution from harmful substances (e.g. heavy metals, persistent organic pollutants, radioactivity, and oil spills) including climate change caused by build-up of GHGs;
- Excessive inputs of nutrients and organic material leading to eutrophication effects;
- Introductions of alien organisms including pathogens and diseases;
- Marked declines and local extinctions of many vulnerable species;
- Degradation of the habitats that provide essential living areas (e.g. feeding, breeding, refuges) for species and communities; and
- Distortions in the characteristic structure, function and integrity of ecosystems which have traditionally made them valuable. Thus, affecting their ability to provide important ecosystem 'goods and services' for humans and so reducing human employment and the viability of coastal communities.

Accordingly, there is a need to actively conserve, and where appropriate, restore the environmental quality (i.e. 'health') of ecosystems (JMM 2003).

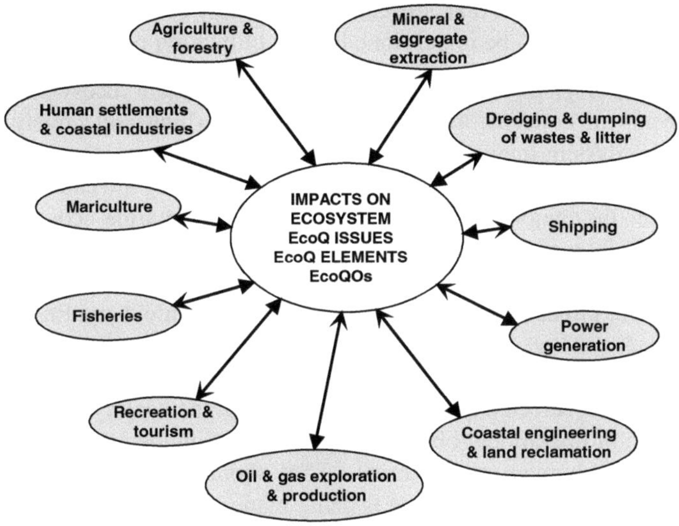

EcoQ Issue/Element
Reference points for commercial fish species
Threatened & declining species
Sea mammals
Seabirds
Fish communities
Benthic communities
Plankton communities
Habitats
Nutrient budgets & production
Oxygen consumption

Fig. 1.2 Many industries and human activities benefit from but also result in pressures that stress marine ecosystems. It is important to identify and rank the human pressures causing problems on ecosystems and to apply prioritized, targeted remedial actions. The EcoQ issues/elements from the 2002 Bergen North Sea Ministerial Declaration are identified

Identifying Key Ecosystem Issues, Setting Objectives for Good Environmental Status and Evaluating Progress

To achieve the vision, and avoid problems related to 'shifting baselines' (Pauly 1995), the EAM requires the development of a coherent system of indicators (NSC 2002; Rice et al. 2005). These should quantify the ecosystem state (i.e. 'health') and severity of human pressures, thereby allowing one to develop and monitor effects-based management actions with the aim of reaching or maintaining 'good environmental status' (GES) (Turnhout et al. 2007; Backer 2008; Heslenfeld and Enserink 2008). The management framework to achieve GES will be most effective when based on objectives, indicators, targets and limits (Rice et al. 2005). In developing EcoQOs and related indicators, it is also important to emphasize that there is a need for *prioritization*, whereby one should devote the greatest effort at regulating the human activities/pressures that are causing the most severe ecosystem impacts. Thus, the vision needs to be underpinned by identification of:

(a) *Strategic ecological issues*. These are the important ecosystem issues or components of concern, the status/condition of which collectively contribute towards ecosystem quality/health as agreed by appropriate management bodies (e.g. HELCOM, OSPAR) for given marine regions. They are chosen to appropriately cover the major components of the particular marine ecosystem that are impacted by human activities. For example, the 2002 Bergen Declaration (NSC 2002) highlighted 10 such issues (Fig. 1.2) which formed the basis for a subsequent OSPAR pilot project focused on the North Sea (Heslenfeld and Enserink 2008);
(b) *Ecological quality objectives (EcoQOs)*: These describe central quality characteristics of a 'healthy' ecosystem (Backer and Leppänen 2008). They specify desired, selected quality characteristics of the healthy marine ecosystem in the form of operational objectives, often based on the analysis of ecosystem properties and human-related threats. They should be related to *indicators* for monitoring whether performance is being made in the right direction or not for achievement of the intended objectives. These indicators are an integral part of the EcoQO system; and
(c) *Indicators*: These provide a measure, an index of measures, or a model that characterizes an ecosystem or one of its key components, providing information displaying environmental change over time. Ideally indicators have three basic functions: (1) simplification; (2) quantification, and (3) communication. Thus, they should be scientifically sound, easily understood, sensitive to the change they are intended to measure/monitor, and measurable and capable of being updated regularly (Rice et al. 2005). They can be roughly divided into two groups based on their use as policy tools: (a) *performance indicators* related directly to effect-based policies, which have quantitative reference values and often target levels to be reached/maintained and/or limit levels which should not be transgressed; and (b) *descriptive indicators* which update information on important ecosystem processes but which do not have quantitative targets set for them (Backer 2008). To be directly usable in a policy or legislative process, a performance indicator must be relatively clearly linked to one or more concrete human pressures (e.g. specific forms of pollution including eutrophication, fishing activities/gear deployment) and so enable the evaluation of the progress of the management actions taken to regulate the causative human activity.

So, to measure 'ecosystem health', several ecological quality issues are chosen, such as status of marine mammals, of seabirds, of oxygen concentrations, of nutrient loads/eutrophication, etc. (NSC 2002). For each issue one or more objectives can be chosen. These objectives each may have one or more indicators for assessing whether the agreed EcoQO is being met. The various indicator indices can be appropriately classified (e.g. good, satisfactory, poor) in terms of their environmental status compared with, for example, a reference condition. The ecosystem is considered to be 'healthy' when all objectives are met (i.e. GES). If this is not so, an appropriate programme of actions and measures (e.g. management/ regulation) may need to be directed at the human activities causing the pressures.

The identification of ecological issues and quantification of relevant EcoQos, with supporting indicators, will vary among systems (e.g. oligotrophic to hypereutrophic areas) or eco-regions (e.g. Baltic Sea, North Sea) and depend on the priority given to various key issues. The setting of EcoQOs should be conducted in an integrated manner, to make certain that the sets are mutually attainable and collectively sufficient to ensure conservation or restoration of the particular ecosystem. It should be a political/societal decision, based on scientific advice, to identify the desired target or threshold (e.g. limit) level of the EcoQO relative to a reference level (e.g. minimally impacted) (Rice et al. 2005).

It is essential to assess the cumulative impacts of multiple human activities on marine ecosystems. Thus, the most important requirement of a fully integrated assessment and management process is to link change in the ecosystem to a manageable human activity, through the pressures which the activities cause (Jennings 2005; Backer 2008). So, it is desirable whenever possible to establish a 'cause and effect' relationship, based on a causal chain analysis, between the human pressures causing serious impacts on one or more components in the ecosystem, so that regulatory measures will result in a substantial decrease in the detrimental ecosystem effects. A DPSIR (Driver – Pressure – State – Impact – Response) analysis represents a useful scheme to organize information about the environment, and has been adopted as a framework and policy tool to identify management options for a range of marine environmental issues (EEA 1995, 2003; FAO 1999).

The main aspects of the DPSIR risk-based assessment scheme (Fig. 1.3) are explained in the bullet points below:

- *Drivers*: The human caused driving forces leading to environmental pressures. Examples are socioeconomic sectoral/industry demands or trends (agriculture, fishing, energy production/utilization, etc.);
- *Pressures*: Examples are emissions/discharges of pollutants, excessive exploitation of resources (land, water, minerals, fish) affecting the state of the environment;

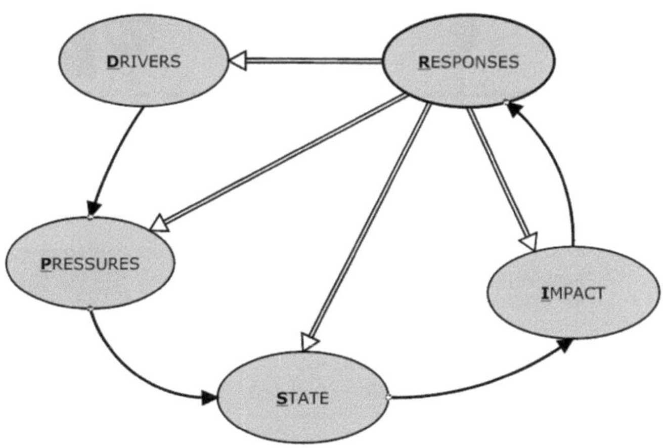

Fig. 1.3 The DPSIR risk-based scheme which provides a framework and policy tool to identify management options for environmental issues

- *State*: Refers to the quality and quantity of environmental media and living resources (air, water, sediments, biota and their habitats) and their consequent ability to provide benefits arising from ecosystem goods and services. The current state = natural state as modified by human pressures;
- *Impacts*: The environmental state (good/poor) may have an impact (beneficial/detrimental) on human health, and the structure, function and integrity of ecosystems which provide amenities and commodities on which human communities depend; and
- *Responses*: Are the efforts by human society (e.g. politicians and decision-makers) to mitigate, adapt to or solve the problems identified by the assessed impacts. Responses include development and application of appropriate environmental policies, management and regulatory measures, including best available technology (BAT) and best environmental practices (BEP), to prevent or eliminate pollution and restore/rehabilitate adversely impacted environmental media/components.

Describing the causal chain from driving forces to impacts and responses is a complex task. Furthermore, it is emphasized that the precautionary principle – as anchored, for example, in the HELCOM statutes – requires measures ('Response' part of the DPSIR scheme) to be taken at source if an adverse effect on the environment is likely, even if a clear cause-and-effect relationship cannot be established.

An illustration of the relationship between various performance-related metrics (reference, target, precautionary, and limit) that define the ecological quality status ranging from 'high' to 'bad' spanning sustainable to unsustainable ecosystem goods and services is provided in Fig. 1.4. The target level(s) (also known as target reference point) identify the EcoQO that management should be trying to maintain

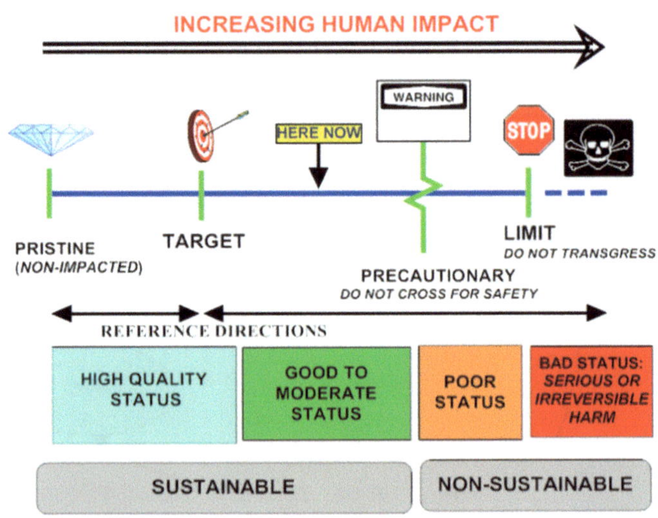

Fig. 1.4 A healthy ecosystem exhibits good accordance with target levels across a suite of ecological quality indicators

or attain with high probability. The limit reference point (LRP) is a value that, if violated (transgressed) is taken as *prima facie* evidence of a conservation concern by which there is unacceptable risk of serious or irreversible harm to the resource or system. LRPs are based on the biology of the stock/species/ecosystem, independent of human socioeconomic considerations. Outside the LRP a state is entered where there is evidence that productivity is seriously compromised, or exploitation is not sustainable, or the stock dynamics are unknown. To account for uncertainty in assessments, precautionary reference points may be identified as a basis for scientific advice, with the intent that management consistent with precautionary reference points (i.e. applying the precautionary approach) should have a high probability of avoiding transgression of a LRP (ICES 2001). In the case of precautionary reference point, the distance between this and the LRP should be correlated with the degree of risk or uncertainty involved. Additionally, many management or regulatory bodies have promoted a simple scoring system of indices, which may integrate or aggregate individual indicators, for comparison purposes. They are frequently expressed as classes of impact (e.g. UNEP 2006: no impact, slight impact, moderate impact, severe impact), environmental quality (EC 2000: high, good, moderate, poor, bad), or coastal condition (US EPA 2004: 'traffic light' approach – good/green, fair/amber, poor/red).

The ecological quality issues and EcoQOs define the topics for which indicators and policy actions have to be developed. Important policy decisions may be taken on general principles, such as an overall commitment to further action, adherence to the precautionary principle, and acceptance of a deviation from a pristine state as a target when the socioeconomic impact (e.g. costs of lost jobs) of reaching pristine status is unacceptable (Backer 2008).

The Importance of Habitat Conservation, Spatial Planning and Establishment of Marine Protected Areas

Based on the major EU policy drivers concerning the marine environment (e.g. WFD, MSFD and Habitats and Birds Directives) one can simplify the management approach used as: (a) managing the impacts of human activities; and (b) protecting identified (e.g. threatened and ecologically important) species and habitats.

Conservation of habitats is essential for protecting the species and biological resources that are dependent on the habitats for their viability (GESAMP 1997; Nilsen et al. 2002). Degradation, fragmentation and eventual habitat loss, together with threats to their characteristic faunal and floral communities, prevail due to encroaching human pressures and climate change (Hopkins 2004; Hopkins et al. 2006; HELCOM 2009). Accordingly, the conservation of habitats is a rapidly growing need with regard to the protection of biodiversity.

The Habitats and Wild Birds Directives, supported by the Natura 2000 initiative, form the cornerstone of the EU's biodiversity conservation strategy and play a major role in achieving the objective of halting biodiversity loss by 2010 (Boedeker and Nordheim 2002). These and other measures, such as the WFD and the MSFD, are of paramount importance. However, despite the important commitments under the directives, their application to the coastal and offshore marine environment in the form of Natura 2000 needs to be substantially accelerated. In accordance with the requirements of the CBD and the Habitats Directive, EU Member States are required to identify, classify and map those species and habitats that are threatened or declining within their national territories. In several cases, Biodiversity Action Plans have been drawn up for the conservation and/or enhancement of priority species or groups of species and their habitats.

Holistic, cross-sectoral marine *spatial planning* is one of the delivery tools of an ecosystem-based management as identified by the EU Maritime Policy and MSFD (Douvere 2008). Marine spatial planning is strategic forward-looking planning for regulating, managing and protecting the marine environment that addresses the multiple, cumulative and potentially conflicting uses of the sea. *Zoning* is a spatial planning tool, acting like a town planning scheme, which allows certain activities to occur in specified areas but recognizes that other incompatible activities should only occur in other specially designated areas, so that zoning provides area-based controls and separates conflicting areas (Day 2000). The zoning process aims at preserving the marine biodiversity, assuring sustainable use of marine resources and an equal and fair use of marine areas, resources and services.

Marine protected areas (MPAs, also called Baltic Sea Protected Areas by HELCOM) are potentially important coastal and offshore areas in which certain uses and/or zones are managed or regulated to conserve the natural resources, biodiversity and human livelihoods. Therefore, MPAs, their conservation goals and spatial needs have to be at the heart of an integrated regional, spatial planning process which aims to minimize conflicts between particular conservation requirements of MPAs and sustainable uses (Christiansen 2009). The ecologically coherent network of well-managed MPAs, to be established by OSPAR and HELCOM in 2010, will consist of the Natura 2000 network comprising protected areas for a selected subset of species and habitats, and additional MPAs according to the specific criteria set by the OSPAR and HELCOM Conventions (JMM 2003).

1.4.3.4 The Framework for Ecosystem-Based Management

Guidance on the application of the EAM in the European marine environment has been provided as input to the development of the European Marine Strategy and the subsequent MSFD (Rice et al. 2005). The conceptual framework for applying ecosystem-based management includes the main components, modified from 5NSC, illustrated in Fig. 1.5, and expanded on in the bullet points below.

The framework includes the following major components:

- Setting agreed management *objectives* for the ecosystem (e.g. establishing EcoQOs);
- Carrying out periodic *monitoring* of the status of key ecosystem components (e.g. selected parameters in sediments, water and biota including key living resources or biodiversity entities and habitats);

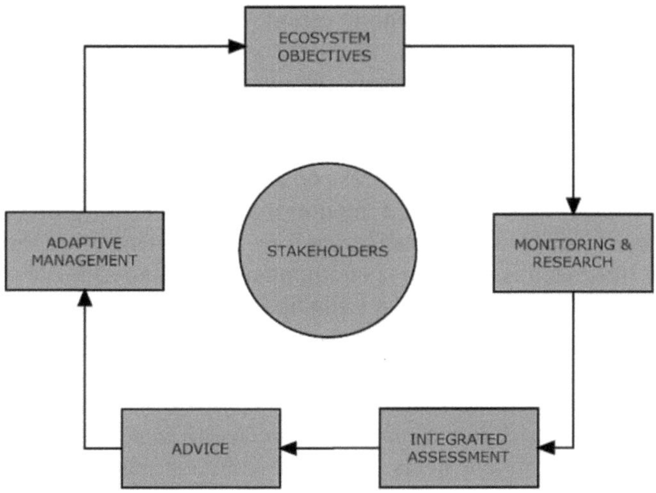

Fig. 1.5 Conceptual framework for an ecosystem approach to management of human activities, involving components identified in the 2002 Bergen Ministerial Declaration on Protection of North Sea

- Conducting *integrated assessments* of the ecosystem based on monitoring data and other relevant knowledge;
- Applying *adaptive management* including interpreting and applying appropriate policies, regulation and ensuring compliance (e.g. control and enforcement) with laws/regulations. One is expected to take actions that are adaptive and precautionary to changing conditions, minimize risk, maximize adaptive capacity (i.e. are anticipatory in addition to being reactive), and are strategic in increasing flexibility via prudent knowledge-based responsive management;
- Providing the best available scientific *advice*, based on evidence, including use of existing knowledge and focused research, for establishing appropriate objectives, improving the monitoring and assessment systems, and proposing new policy and management measures; and
- Inclusion of all relevant *stakeholders* to ensure that appropriate wide-reaching consultation, dialogue and eventually consensus are involved for reflecting societal views in the various parts of the decision-making process.

The *framework* (Fig. 1.5) starts with actions to generate information concerning the ecosystem and interacting human activities. This is done by *monitoring to assess the state of the ecosystem* and through *research*, giving insight into relationships, interactions, and processes guiding the ecosystem. Together this information feeds the central line and dominating part of the framework, the *integrated assessment*. The integrated assessment is subject to the management *objectives* that are established for the specific marine ecosystem. *Comparison of the outcome of the integrated assessment compared with the objectives will result in scientific advice to the management regarding what measures should be applied to achieve the objectives set.* This advice is used by managers and policymakers to set up a *management regime* for the coming period. The *effect* of this new management regime is measured through *monitoring*. The whole process is repeated periodically. In the real world there are many *interactions between the parties* involved and this *communication* forms an important aspect of the ecosystem approach.

To support *integrated assessments*, *monitoring programmes* provide updated information on the status and trends of components of the ecosystem. There is a need to move towards integrated monitoring in an ecosystem context. The *integrated assessment* is a major factor that forces other elements of the framework to deal with integrated issues, and is an important scientific element of an ecosystem approach. Assessment at regular intervals of the status of the marine environment, its ecosystem, and the degree of anthropogenic influence on the ecosystem, forms the basis for scientific advice to managers.

For *research and monitoring*, this can be interpreted as *multidisciplinary research and integrated monitoring* where, at least, *data exchange* and *quality assurance* between different fields of work is common practice. As noted previously, the process to *define specific objectives* for the management of marine ecosystems *is a major challenge* in implementing an ecosystem approach. There is interaction between *scientific knowledge*, *socioeconomic forces* (e.g. human employment and livelihood) and national and international *policies* ending up in a political decision-making process. The attention to *socioeconomics* and the need to involve *stakeholders* in the decision-making process reflects the fact that diverse human activities benefit from and result in impacts on ecosystems. Ecosystem management can only be achieved by the *regulation* (i.e. control and enforcement) *of human activities*, which is the overarching reason to include stakeholder representation. *Communication and cooperation* (e.g. confidence building, trust) within and between various sectoral groups such as scientists (e.g. natural, economic, social and political disciplines), stakeholders (e.g. industries, environmental NGOs) and managers (e.g. fisheries, environment ministries and commissions) is essential to reach consensus on any new objective.

1.5 How Do Climate Change Policy, Management and Research Get Integrated into This?

The two main ways to tackle climate change are by:

- *Mitigation* – taking actions to slow the speed and amount of climate change by reducing greenhouse gas emissions and increasing their absorption by enhancing 'sinks' (BEAC 2005); and
- *Adaptation* – taking actions concerning 'adjustment' of natural or human systems in response to actual or expected climatic stimuli or their effects, which moderates harm or exploits beneficial opportunities (IPCC 2007b). Adaptation covers actions to attempt to limit adverse impacts by becoming more resilient to the climate changes that will occur while mankind pursues mitigation.

1.5.1 Mitigation

There is substantial economic potential for the mitigation of global GHG emissions over the coming decades, which could offset the projected growth of global emissions or reduce emissions below current levels (IPCC 2007a, b).

Mitigation involves each of the marine industries, in addition to terrestrial industries, finding ways to reduce their GHG emissions. This involves, among other possibilities, application, and further development of BAT and BET.

The marine industries can contribute by:

(a) *Increasing their energy efficiency*, and
(b) *Using renewable energy sources* such as solar power, wind power, geothermal and water power, and even nuclear energy.

As an example, one can consider the case of the *fisheries sector*. Although this sector accounts for only about 1.2% of global oil consumption, the global average of fuel used per unit weight of catch landed is about 0.45 kg fuel per kg landings (640 L of fuel per tonne landings), but it can reach up to 6 kg fuel per kg landings in some fisheries such as bottom trawling for prawns or beam-trawling for flat-fish (Pimentel et al. 2007). Most of this fuel consumption originates from the sea-going harvesting sub-sector rather than the land-based food processing sub-sector. There have been few incentives to change in the direction of greater fuel efficiency in the various areas of shipping (e.g. fisheries, transport) due to little if any tax on commercial marine fuels, although fuel price escalation has helped.

Improving fuel efficiency by, for example, the development of new technologies and greater use or even improved design of passive, more target-specific fishing gears is becoming increasingly desirable, in response to the cost of wild-fish capture escalating through rising fuel prices. Associated with RTD into better gear is technical research into enhanced ship hull and propulsion design, lowering the carbon/emissions footprint of the current fleets and making optimal use of environmentally friendly energy sources.

Although ships emit less CO_2 than other transport modes (tonnage per km sailed), which has given them a green image previously, this image is fading away. By 2020 ships will be major contributors to pollution and climate warming by emitting as much SOx and NOx in EU regional seas as all land based sources in the EU put together.[4]

1.5.2 Adaptation

Adaptive measures are needed to minimize the adverse impacts of climate change (e.g. global warming) while we carry out the mitigation measures which will take several decades to gain tangible effect (IPCC 2007b; EC 2009c). There are

[4] http://www.flad.pt/documentos/1227109470G4fJQ9kv7Mi01DK7.pdf

numerous ways that adaptation can be promoted at the general levels of good practice and the specific levels which apply to particular human sectors or industries. The latter are too detailed and abundant to be considered here, so the former will form the current focus. The generic considerations are intended to be illustrative and not definitive.

Some marine ecosystems and their biota may suffer and some may benefit as a result of climate change (EFARO 2010). Accordingly, management must prudently and optimally respond to climate change by adaptively reflecting interactions with relevant human pressures, and focus on the changing status and trends of biological resources (e.g. re-distributions and migrations of both 'old' and 'new' biological resources, including changes in reproduction and recruitment, growth and productivity, food availability, multispecies interactions and food-webs) and their habitats (e.g. carrying capacity for key stocks/biota including hydrodynamic and oceanographic environment affecting their viability). In particular, adaptation involves tackling the adverse impacts of global warming by becoming more resilient to the climate changes that will occur before mitigation measures prove effective.

Environmental management policies should better incorporate and integrate the effects of climate change and variability in adaptive actions. The main point to recognize is that climate change interacts with other multiple stressors of the marine environment (e.g. excessive 'extractive' harvesting of living marine resources, pollution including eutrophication, and habitat degradation). Thus, there is a need to reduce or limit human caused stressors. Accordingly, it is essential that good environmental management is followed via applying the EAM to ensure that ecosystems are healthy and resilient. In particular, resilience to climate change effects will be facilitated by ensuring that human activities and their impacts on ecosystem components are kept well within 'precautionary limits'. Especially, 'good environmental status' must be achieved and maintained regarding the WFD and the MSFD.

In areas such as fisheries and aquaculture, for example, adaptations to climate should be included in establishing harvesting levels, rules and practices, and developing prudent adaptive strategies. The CFP must urgently reduce fishing pressure to sustainable levels which maintain fish stocks within safe biological limits. Sustainable fishing therefore has to replace overfishing which has rendered marine ecosystems more vulnerable to climate change and thus less capable of adapting (EC 2009a). In addition to increasing their resilience to climate change and decreasing their variability, this may facilitate achieving two other desirable goals, viz. achieving longer term sustainable yields from such resources and reducing GHG emissions in their harvesting (Brander 2008).

The WFD and MSFD identify changing background conditions as part of the assessment of reference ecological/environmental conditions. The definition of reference condition is type-specific, and therefore requires an analysis of natural variability, with periodic reassessment of natural variability. Thus, it allows for a re-appraisal of reference conditions (e.g. target-, limit- and precautionary-levels) which incorporate the impact of climate change on the ecosystem. The WFD's River Basin Management Plans due in 2009 are charged with taking into account the impacts of climate change and the next generation of plans due in 2015 should strive to be fully climate-proofed. Climate change must also be properly integrated in the implementation of the MSFD which requires the achievement of good environmental status of the EU's marine waters by 2020. Full implementation of this directive will help increase resilience in the marine environment and facilitate adaptation efforts. A more coherent and integrated approach to maritime and coastal planning and management is also necessary. The Integrated Maritime Policy will provide a comprehensive framework to integrate adaptation efforts coherently into sectoral and specific policies and measures. Efforts must be stepped up to ensure that the provisions in the 2002 Integrated Coastal Zone Management (ICZM) Recommendation (EC 2002) involving EU Member States are fully respected and strengthened. The follow-up to the 2008 European Commission Communication entitled Roadmap for Maritime Spatial Planning will incorporate adaptation to climate change in maritime and coastal management (EC 2008b).

Climate change must be integrated into sectoral policy, planning and related activities. Effective use must be made of a ground level action plan to demonstrate implementation of synergies between 'agreements' as proposed by COP 9 of the UNFCCC (cf. UN 2004). Consideration of the climate change adaptation challenge is basically an integrative challenge across sectors and policies.

Adaptations in marine environmental management systems must be well-informed and flexibly formulated to take account of uncertainty in future climate change and its impacts. So, a close link between science and policy is vital.

1.6 The Importance of Linking Science and Policy

This section examines the important interactions between science and policy, including challenges concerning the organization and funding of relevant research, and the ways that scientific scenario building can be used to envisage potential environmental developments and policy options.

1.6.1 Scientific Support for Policy

As emphasized earlier, an important overall objective of scientific research connected with developing and implementing the EAM is (a) differentiating between naturally and human induced pressures, (b) discovering their interactions and combined effects, and (c) understanding the related consequences (e.g. impacts, costs and benefits) of changes in the ecosystem for human communities. Related to this is the important aim of facilitating the dissemination of the scientific outputs for use by the wider public including national and international advisory bodies and executive authorities. A conceptual sketch of the supply chain running from scientific research to regulation and management is shown in Fig. 1.6.

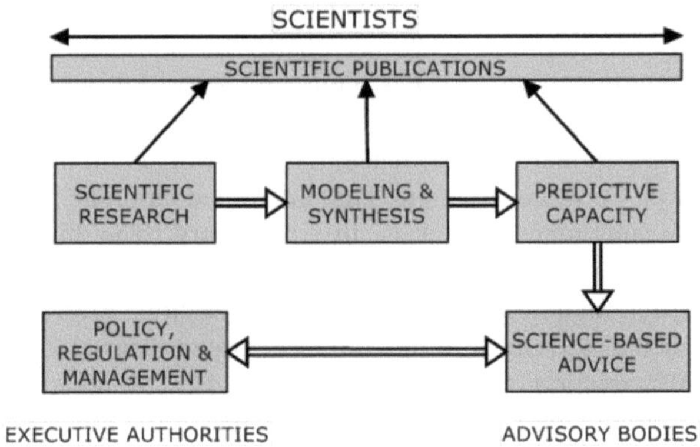

Fig. 1.6 The supply chain linking science and regulation and management

Scientific support is crucial for policy development and implementation, including achieving effective and adaptive management, regulatory and mitigation/remedial measures that are needed to implement the EAM. Research has a vital role in reducing the extent of the uncertainty and risk. The best available research and scientific evidence is required to underpin the required knowledge inputs for the entire management and regulatory process. This knowledge is essential in the design, implementation and evaluation of policies, and also to achieve effective mitigation measures for harmful human activities. For example, sound ecosystem-based management must be founded on widely and credibly integrating scientific data and knowledge that involves the interactions between the ecosystem and environmental components and the cumulative impacts of different human activities (Hopkins et al. 2006). As the EAM is an evolutionary and adaptive process, the science community has a vital function in producing the RTD needed to provide the best available, politically impartial, scientifically objective knowledge and advice required by the EAM during its evolutionary application.

Thus, *scientific expertise is a crucial element in the design, implementation and evaluation of the efficacy of public policies, as well as for finding ways to remedy and avoid detrimental effects of human activities.* Environmental and socioeconomic sustainability emphasizes the need for knowledge integration and mobilization, dialogue and consultation with a view to reaching accord, and wider collaboration to address 'real world' problems (UN 2002). Policy should be supported by excellence in marine science and technology, and integration across all scientific themes and areas is essential for maximizing the benefits from marine research (Borg 2005). The sustainability issues are diverse and complex and so the science and policy interface is shifting from individualistic discipline/issue and segregated sector-related science and management to more holistic (i.e. integrated) and utility-focused. This must intersect sectoral research, management and governance boundaries.

Today research activities and the funding of gaps in knowledge are shaped and determined not only by scientists but also by other actors in society who have various intellectual and social backgrounds, motivations and user requirements

(Hopkins et al. 2006). This produces an increasing demand and supply for more timely and practical value-for-money knowledge that correspondingly fosters closer, more interactive engagement between researchers and those who fund and apply research. The outcomes of integrated environmental assessments of current and future developments will lead to holistic scientific information and advice, lessons learnt and best practices to form the basis for management options. All these issues must be widely communicated in a more consistent, coherent and convincing manner in order to satisfy societal choice and acceptance criteria.

Evolution of the European research organization is essential in order to optimally support the capacity of the science community to maximize both the quality and quantity of the research outputs from the specific and cross-cutting priorities. These should enhance the positive trends related to fulfilling the aims of the European Research Area. Actions are needed to facilitate the following essentials:

- An appropriate 'mix' of sources (e.g. private, public, national, European) and allocations (e.g. basic and applied research) regarding research planning and funding, including new, more varied and collaborative sectoral partnerships (e.g. public – private, academia – industry);
- Enhanced funding and access to modern research infrastructures including more effective procurement planning and sharing strategies;
- Novel education and high-quality training to produce research scientists, technicians and research managers with a wider, interdisciplinary and inter-sectoral knowledge of ecosystem-based science, advice and management. Efforts must tackle the limited interdisciplinary collaboration, and lack of understanding, between natural, technological, economic, social, and political sciences;
- Lack of inclusion and funding of the social, economic and political sciences severely limits integration of the 'human dimension' into research that is traditionally dominated by the natural sciences; and
- New approaches are required for networking, knowledge dissemination and the sharing of data between different sectors and communicating complex research insights, to effectively build confidence and credibility among diverse stakeholder groups.

1.6.2 Scenarios: Ways to Envisage Developments and Policy Options

Scenarios may be used to consider how recent past, current (today) and emerging ecosystem changes may become future reality. They can take the form of narratives (stories) of imagined 'futures', providing an account or synopsis of various projected trajectories, from today's baseline (*status quo*) status for the system (e.g. fisheries and its component subsystems). They are useful for analyzing sources, patterns and causes (e.g. in a DPSIR context) of change, identifying vital issues (core challenges), and for mapping alternative possible futures/'outlooks' (*what if?*).

So, it is possible to consider a range of plausible but potentially equally relevant development alternatives/options including optimistic or desirable and problematic or undesirable futures. Scenarios should not come singly as a forecast would, but in sets of alternatives/options. They are able to draw on qualitative and quantitative knowledge about the driving forces (i.e. drivers) and pressures acting on various component subsystems making up the overall system.

The main purpose of scenario building is to support strategic planning and enable decision-makers to explore the major, alternative futures, thereby clarifying intervention options (responses) and their possible consequences. Based on identification of the pertinent challenges, far-sighted research agendas can be established to promote the desirable (positive) trends or reduce the undesirable (negative) impacts. Thus, scenarios represent an important tool for linking science and policy.

It is prudent to incorporate scientifically-based scenarios and models of climate change and variability, and human encroachment into management and decision-making. Scenarios and models are needed for decisions involving the conservation and utilization of living resources, and the planning and location of human constructions and settlements. Such an approach provides greater prudence and resilience in tackling current, emerging and future challenges.

In recent years, there is a been a substantial increase in efforts directed at applying scenarios in a Baltic Sea context coupled with climate – ecosystem – socioeconomic models and decision support systems or tools, for example in the Mare and BONUS EEIG research projects. Such initiatives include Baltic Nest, and the ECOSUPPORT, Baltic-C and AMBER projects.

An example of a Baltic Sea scenario, for consideration, is: *Under anticipated climate warming of the Baltic Sea and its effects on water exchange/flushing of the system with consequences for temperature, salinity and oxygen changes, plus management measures instigated by the HELCOM Baltic Sea Action Plan to reduce the eutrophication, what may happen to recruitment, growth and productivity of cod, herring and sprat stocks, and what are the prudent longer term options/ strategies for sustainable fisheries management?*

Acknowledgements This paper is based on a lecture given by the author at the summer 2009 Nordic Marine Academy Ph.D. course 'Climate impacts on the Baltic Sea – from science to policy' on the island of Bornholm. The initiative and encouragement of Brian MacKenzie, Keith Brander, Anders Omstedt and Marcus Reckermann is gratefully recognized in coordinating the course and helping to focus this lecture. In particular, I thank the referees for their advice in improving the quality of the publication.

References

Backer H (2008) Indicators and scientific knowledge in regional Baltic Sea environmental policy. ICES J Mar Sci 65:1398–1401

Backer H, Leppänen J-M (2008) HELCOM ecological objectives for an ecosystem approach: the process of defining good ecological status of the Baltic Sea. Aquat Conserv 18:321–334

BBC (1993) BBC English dictionary: a dictionary for the world. HarperCollins, London
BEAC (2005) Arctic climate: policy measures relevant for the Barents Region. Committee of Senior Officials working document, final Doc. BEAC.CSO.2005.31. 17 Oct 2005. Barents Euro-Arctic Council, 12 pp
Boedeker D, von Nordheim H (eds) (2002) Application of NATURA 2000 in the marine environment. German Federal Agency for Nature Conservation. BfN – Skriften 56, 105 pp
Borg J (2005) Speech 'How can marine research contribute to the Maritime Policy?', given by Commissioner Joe Borg at the European parliament conference on the future of European marine scientific research, Brussels, 17 Oct 2005
Brander K (2008) Tackling the old familiar problems of pollution, habitat alteration and overfishing will help with adapting to climate change. Mar Pollut Bull 56(12):1957–1958
Brundtland GH (ed) (1987) Our common future: the world commission on environment and development. Oxford University Press, Oxford
CBD (1992) Convention on biological diversity. United Nations Environment Programme, Nairobi, June 1992
CBD (1998) The 12 Malawi principles for the ecosystem approach, emanating from the workshop on the ecosystem approach (Lilongwe, 26–28 Jan 1998). Presented at the fourth meeting of the conference of the parties to the convention on biological diversity, Bratislava, 4–15 May 1998. UNEP/CBD/COP/4/Inf.9
CBD (2003) Report of the Ad Hoc technical expert group on biological diversity and climate change. Interlinkages between biological diversity and climate change: advice on the integration of biodiversity considerations into the implementation of the United Nations Framework Convention on Climate Change and its Kyoto Protocol. CBD technical series no. 10. Secretariat of the Convention on Biological Diversity, Montreal
Christiansen S (2009) Towards good ecosystem health – a network of marine protected areas for the North Sea. WWF Germany, Frankfurt am Main
Costanza R, d'Arge R, de Groot R, Farber S, Grasso M, Hannon B, Naeem S, Limburg K, Paruelo J, O'Neill RV, Raskin R, Sutton P, van den Belt M (1997) The value of the world's ecosystem services and natural capital. Nature 387:253–260
Day J (2000) Zoning-lessons from the Great Barrier Reef Marine Park. Ocean Coast Manage 45:139–156
de Sadeleer N (ed) (2007) Implementing the precautionary principle: approaches from the Nordic countries, EU and USA. Earthscan, London, 396 pp
Dinter WP (2001) Biogeography of the OSPAR Maritime area – a synopsis of the biogeographical distribution patterns described for the North–East Atlantic. Federal Agency for Nature Conservation, Bonn, 167 pp
Douvere F (2008) The importance of marine spatial planning in advancing ecosystem-based sea use management. Mar Policy 32(5):762–771
EC (1987) Single European Act signed in Luxembourg on 17 February 1986 and The Hague on 28 February 1986. Official Journal of the European Communities L 169/1 of 29.6.1987
EC (2000) Directive 2000/60/EC of the European Parliament and of the Council of 23 October 2000 establishing a framework for community action in the field of water policy ('Water Framework Directive'). Official Journal of the European Communities L 327/1 of 22.12.2000
EC (2002) Recommendation of the European Parliament and of the Council of 30 May 2002 concerning the implementation of Integrated Coastal Zone Management in Europe. Official Journal of the European Communities L 148/24 of 6.6.2002
EC (2006) Towards a future Maritime Policy for the Union: a European vision for the oceans and seas. Communication from the commission to the council, the European Parliament, the European Economic and Social Committee and the Committee of the Regions. COM (2006) 275 final of 7.6.2006 [Maritime Policy Green Paper]
EC (2007) An integrated Maritime policy for the European Union. Communication from the Commission to the European Parliament, the Council, the European Economic and Social Committee and the Committee of the Regions. COM (2007) 575 final of 10.10.2007 [Maritime Policy Blue Book]

EC (2008a) Directive 2008/56/EC of the European Parliament and of the Council of 17 June 2008 establishing a framework for community action in the field of marine environmental policy (Marine Strategy Framework Directive). Official Journal of the European Union L 164/19 of 25.6.2008
EC (2008b) Communication from the commission. Roadmap for maritime spatial planning: achieving common principles in the EU. COM (2008) 791 final of 25.11.2008
EC (2009a) Green paper – reform of the common fisheries policy. COM (2009) 163 final of 22.4.2009
EC (2009b) Council Regulation (EC) No. 439/2009 of 23 March 2009 concerning the conclusion of the agreement between the European Community and the Government of the Russian Federation on cooperation in fisheries and the conservation of the living marine resources in the Baltic Sea. Official Journal of the European Union L 129/1 of 28.5.2009
EC (2009c) White paper – adapting to climate change: towards a European framework for action. COM (2009) 147 final of 1.4.2009
EC (2010) Commission Decision of 1 September 2010 on criteria and methodological standards on good environmental status of marine waters. Official Journal of the European Union L 232/14 of 2.9.2010
EEA (1995) Europe's environment – the Dobris assessment. European Environment Agency, Copenhagen, 712 pp
EEA (2003) Europe's environment – the third assessment. Environmental Assessment Report No. 10. European Environment Agency, Copenhagen
EEC (1979) Council Directive 79/409/EEC of 2 April 1979 on the conservation of wild birds. Official Journal L 103 of 25.4.1979
EEC (1992) Council Directive 92/43EEC of 21 May 1992 on the conservation of natural habitats and of wild fauna and flora. Official Journal L 206 of 22.7.1992
EFARO (2010) EFARO's view on the future of European fisheries and aquaculture research. European Fisheries and Aquaculture Research Organizations, IJmuiden, 28 pp
Ehlers P (1994) The Baltic Sea area: convention on the protection of the marine environment of the Baltic Sea area (Helsinki Convention) of 1974 and the revised convention of 1992. Mar Pollut Bull 29:617–621
FAO (1995) Code of conduct for responsible fisheries. FAO, Rome, 41 pp
FAO (1996) Precautionary approach to capture fisheries and species introductions. Elaborated by the technical consultation on the precautionary approach to capture fisheries (including species introductions), Lysekil, 6–13 June 1995. FAO Technical Guidelines for Responsible Fisheries No. 2, Rome, 54 pp
FAO (1999) Indicators for sustainable development of marine capture fisheries. FAO Technical Guidelines for Responsible Fisheries. No. 8. Fishery Resources Division. FAO, Rome, 68 pp
GESAMP (1997) Marine biodiversity: patterns, threats and conservation needs. Rep Stud GESAMP 62, 24 pp
GIWA (2005) Global International Waters Assessment – Baltic Sea GIWA regional Assessment 17. UNEP, Kalmar, 69 pp
Heileman S, Thulin J (2009) XII-35 Baltic Sea LME. In: Sherman K, Hempel G (eds) The UNEP large marine ecosystem report: a perspective on changing conditions in LMEs of the world's regional seas. UNEP Regional Seas Report and Studies No. 182. United Nations Environment Programme, Nairobi
HELCOM (1992) Convention for the protection of the marine environment of the Baltic Sea area, 1992 (Helsinki Convention). http://www.helcom.fi/stc/files/Convention/Conv0704.pdf. Accessed 21 Oct 2001
HELCOM (2003) The Baltic marine environment 1999–2002. Balt Sea Environ Proc 87, 48 pp
HELCOM (2007a) Climate change in the Baltic Sea area – HELCOM thematic assessment in 2007. Balt Sea Environ Proc 111, 49 pp
HELCOM (2007b) HELCOM Baltic Sea action plan, adopted on 15 November 2007 in Krakow, Poland by the HELCOM Extraordinary Ministerial Meeting. http://www.helcom.fi/stc/files/BSAP/BSAP_Final.pdf. Accessed 21 Oct 2010

HELCOM (2009) Biodiversity in the Baltic Sea – an integrated thematic assessment on biodiversity and nature conservation in the Baltic Sea. Balt Sea Environ Proc 116B, 188 pp
HELCOM (2010) Ecosystem health of the Baltic Sea 2003–2007: HELCOM initial holistic assessment. Balt Sea Environ Proc 122, 63 pp
Heslenfeld P, Enserink EL (2008) OSPAR ecological quality objectives: the utility of health indicators for the North Sea. ICES J Mar Sci 65:1392–1397
Holden M (1994) The Common Fisheries Policy: Origin, evaluation and future. Fishing News Books, Oxford, 274 pp
Hopkins CCE (2004) Biodiversity assessment and threats analysis for the WWF global 200 EcoRegion 'North–East Atlantic Shelf'. WWF Germany, Frankfurt am Main, 108 pp
Hopkins CCE, Thulin J, Sutinen JG, Kononen K, Snoeijs P, Johansson S, Nalbach H-O (2006) BONUS-169 Baltic Sea science plan and implementation strategy. BONUS Publication No. 5. Academy of Finland, Helsinki
ICES (1964) Convention for the International Council for the Exploration of the Sea, Copenhagen, 12 Sep 1964. http://www.ices.dk/aboutus/convention.asp. Accessed 21 Oct 2010
ICES (2001) Report of the ICES Advisory Committee on Ecosystems 2001. ICES Cooperative Research Report No. 249
ICES (2003) Environmental status of the European seas. A quality status report prepared by the International Council for the Exploration of the Sea for the German Federal Ministry for the Environment. Nature Conservation and Nuclear Safety, Copenhagen, 75 pp
IMM (1997) Statement of conclusions from the Intermediate Ministerial Meeting on the integration of fisheries and environmental issues, Bergen, 13–14 Mar 1997
IPCC (2001) Climate change 2001: Synthesis Report. A contribution of working groups I, II, and III to the Third Assessment Report of the Intergovernmental Panel on Climate Change [Watson RT, the Core Writing Team (eds)]. Cambridge University Press, Cambridge, 398 pp
IPCC (2007a) Climate change 2007: Synthesis Report. Contribution of Working Groups I, II and III to the Fourth Assessment Report of the Intergovernmental Panel on Climate Change [Core Writing Team, Pachauri RK, Reisinger A (eds)]. IPCC, Geneva, 104 pp
IPCC (2007b) Climate change 2007: Impacts, Adaptation and Vulnerability. Contribution of Working Group II to the Third Assessment Report of the Intergovernmental Panel on Climate Change [Parry ML, Canziani OF, Palutikof JP, van der Linden PJ, Hanson CE (eds)]. Cambridge University Press, Cambridge, 1000 pp
Jennings S (2005) Indicators to support an ecosystem approach to fisheries. Fish Fish 6:212–232
Jennings S, Brander K (2010) Predicting the effects of climate variation and change on marine communities and the consequences for fisheries. J Mar Syst 79:418–426
JMM (2003) Statement on the ecosystem approach to the management of human activities: Towards an ecosystem approach to the management of human activities. Annex 5. First joint meeting of the Helsinki and OSPAR Commissions (JMM), Bremen, 25–26 June 2003
Longhurst AR (1998) Ecological geography of the sea. Academic Press, San Diego, 398 pp
Lubchenco J (1994) The scientific basis of ecosystem management: framing the context, language and goals, pp 33–39. In: Committee on environment and public works, United States Senate, ecosystem management: status and potential. Proceedings of a workshop by the congressional research service, 24–25 March 1994. 103rd congress, 2nd session. U.S. Government Printing Office, Washington, DC
Nilsen H-G, Aarefjord H, Øverland S, Rukke J (2002) Progress report. Fifth international conference on the protection of the North Sea, Bergen, 20–21 Mar 2002. Norwegian Ministry of the Environment. ISBN – 82-457-0353-2
NOAA (2005) Background paper on large marine ecosystems and approach to assessment and management. Protection of the arctic marine environment (PAME) meeting, Copenhagen, 22–23 Feb 2005. NOAA – Fisheries, Narragansett, 31 pp
NSC (1995) Esbjerg declaration. 4th international conference on the protection of the North Sea, Esbjerg, 8–9 June 1995. Ministry of Environment and Energy, Danish Environmental Protection Agency, Copenhagen, 142 pp

NSC (2002) Bergen declaration. Fifth international conference on the protection of the North Sea, Bergen, 20–21 Mar 2002. Norwegian Ministry of the Environment, 170 pp
OSPAR (2010) Quality status report 2010. OSPAR Commission, London, 176 pp
Pauly D (1995) Anecdotes and the shifting baseline syndrome of fisheries. Trends Ecol Evol 10 (10):430
Perry RI, Cury P, Brander K, Jennings S, Möllmann C, Planque B (2010) Sensitivity of marine systems to climate and fishing: concepts, issues and management responses. J Mar Syst 79:427–435
Pimentel D, Shanks RE, Rylander JC (2007) Energy use in fish and aquaculture production. In: Pimentel D, Pimentel M (eds) Food, energy, and society. CRC Press, Boca Raton, pp 77–95
Planque B, Fromentin J-M, Cury P, Drinkwater KF, Jennings S, Perry IR, Kifani S (2010) How does fishing alter marine populations and the ecosystem sensitivity to climate? J Mar Syst 79:403–417
Rice JR, Trujillo V, Jennings S, Hylland K, Hagström O, Astudillo A, Nørrevang Jensen J (2005) Guidance on the application of the ecosystem approach to management of human activities in the European marine environment. ICES Cooperative Research Report No. 273, 22 pp
Rozwadowski HM (2002) The sea knows no boundaries: a century of marine science under ICES. University of Washington Press, Seattle/London, 410 pp
Sherman K (1994) Sustainability, biomass yields, and health of coastal ecosystems: an ecological perspective. Mar Ecol Prog Ser 112:277–301
Sherman K, Duda A (1999) An ecosystem approach to global assessment and management of coastal waters. Mar Ecol Prog Ser 190:271–287
Sissenwine M, Symes D (2007) Reflections on the common fisheries policy. Report to the General Directorate for Fisheries and Maritime Affairs of the European Commission. European Commission, Brussels, 75 pp
Suárez de Vivero J, Rodriguez Mateos JC, Florido del Corral D (2009) Geopolitical factors of maritime policies and marine spatial planning: state, regions, and geographical planning scope. Mar Policy 33(4):624–634
Thulin J (2009) The recovery and sustainability of the Baltic Sea large marine ecosystem. In: Sherman K, Aquarone MC, Adams S (eds) Sustaining the world's large marine ecosystems. IUCN, Gland
Turnhout E, Hisschermölle M, Eisjackers H (2007) Ecological indicators between the two fires of science and policy. Ecol Indic 7:215–228
UN (1997) Kyoto Protocol to the United Nations Framework Convention on Climate Change (UNFCCC), Adopted by consensus at the third session of the conference of the parties (COP3) in December 1997 on 11 December 1997 in Kyoto. Document FCCC/CP/1997/7/Add.1 25 Mar 1998
UN (2002) Report of the world summit on sustainable development. Johannesburg 26 Aug to 4 Sep. 2002. Report No. A/CONF.199/20
UN (2004) Report of the conference of the parties on its ninth session (COP 9), 1–2 Dec 2003, Milan. United Nations Framework Convention on Climate Change (UNFCCC). Addendum part two: actions taken by the conference of the parties at its ninth session. Document FCCC/CP/2003/6/Add.1
UN (2007) UN climate change conference, 13–15 Dec 2007, Bali. United Nations Framework Convention on Climate Change (UNFCCC). Thirteenth session of the conference of the parties (COP 13) document FCCC/CP/2007/6/Add.1
UN (2009) UN climate change conference, 7–19 Dec 2009, Copenhagen. Fifteenth session of the conference of the parties (COP 15). Actions taken: document FCCC/CP/2009/11/Add.1
UNCCD (1994) Elaboration of an international convention to combat desertification in countries experiencing serious drought and/or desertification, particularly in Africa. A/AC.241/27. 12 Sep. 1994. United Nations, New York, 58 pp
UNCED (1992) United Nations conference on environment and development. Rio declaration of the earth summit agenda 21, Rio de Janeiro. United Nations, New York

UNEP (2006) Challenges to international waters – regional assessments in a global perspective. United Nations Environment Programme, Nairobi, 120 pp

UNFCCC (1992) United Nations Framework Convention on Climate Change. United Nations, New York, 24 pp

US EPA (2004) National coastal condition report II 2005. EPA-620/R-03/002. U.S. Environmental Protection Agency. Office of Research and Development and Office of Water, Washington, DC, 311 pp

Chapter 2
HELCOM Baltic Sea Action Plan: An Ecosystem Approach to the Management of Human Activities

Minna Pyhälä

Abbreviations

BSAP	HELCOM Baltic Sea Action Plan
EC	European Community
EQR	Ecological Quality Ratio
EU	European Union
HELCOM	Helsinki Commission
IMO	International Maritime Organization
MSFD	Marine Strategy Framework Directive
NECD	EU National Emission Ceiling Directive
NEFCO	Nordic Environmental Finance Corporation
NIB	Nordic Investment Bank
OSPAR	Commission for the Protection of the Marine Environment of the North-East Atlantic
WFD	EU Water Framework Directive

2.1 Introduction

The Helsinki Commission (or HELCOM for short) is an intergovernmental organization comprising the Baltic Sea coastal states and the European Union (EU). HELCOM has been working since the early 1970s to improve the status of the

The author works as Assisting Professional Secretary at the HELCOM Secretariat. Some of the views presented in this chapter are those of the author and may not necessarily reflect those of the Helsinki Commission.

M. Pyhälä (✉)
Helsinki Commission, Katajanokanlaituri 6B, 00160 Helsinki, Finland
e-mail: minna.pyhala@helcom.fi

M. Reckermann et al. (eds.), *Climate Impacts on the Baltic Sea: From Science to Policy*, Springer Earth System Sciences, DOI 10.1007/978-3-642-25728-5_2,

Baltic Sea marine environment. After several decades of addressing threats to the marine environment on a sector-by-sector basis, in the early 2000s HELCOM began working towards adopting a more holistic approach to protecting the Baltic Sea environment. This work began with the adoption of a common vision of the application of the ecosystem approach to managing human activities together with the Commission for the Protection of the Marine Environment of the North-East Atlantic (OSPAR) at a joint ministerial meeting in 2003 (JMM 2003) and culminated in the adoption, in 2007, of the HELCOM Baltic Sea Action Plan (BSAP, HELCOM 2007a).

This chapter gives an introduction to the work of HELCOM, focusing mainly on the BSAP and how it makes use of scientific information to support decision-making and the adoption of cost-effective measures to reduce pressures to the Baltic Sea marine environment.

Eutrophication is one of the most significant environmental problems faced by the Baltic Sea. Because a substantial amount of eutrophication related data (including long time-series) exist, Baltic Sea scientists have been able to develop indicator based assessment tools that support decision-making and allow for monitoring progress towards reaching jointly agreed environmental targets. The case of eutrophication is thus used here as an example to illustrate a full management cycle based on the ecosystem approach.

Climate-related ecosystem changes are already taking place in the Baltic Sea (HELCOM 2007b). In addition to new threats to the marine environment, many existing problems, such as eutrophication, are likely to be exacerbated by the added stress of climate change. It is important to consider the marine ecosystem as a whole and modify anthropogenic activities to reduce their impacts as further information becomes available. To ensure appropriate measures are put into place, in addition to political will and financial resources, it is important to improve the linkages between science and policy.

2.2 The Helsinki Convention: Protecting the Baltic Sea Marine Environment

The Convention for the Protection of the Marine Environment of the Baltic Sea Area (Helsinki Convention) was signed in 1974 by the Ministers of Environment from the (then) seven Baltic Sea coastal states[1] (HELCOM 1974). The Convention was developed during an era of increasing environmental awareness and as a

[1] The contracting parties of the 1974 Helsinki Convention were Denmark, Finland, German Democratic Republic, Federal Republic of Germany, Polish Peoples Republic, Union of Soviet Socialist Republics and Sweden. In 1992, after the fall of communism in the region, the Helsinki Convention was revised and signed by Denmark, Estonia, European Economic Community (now European Union), Finland, Germany, Latvia, Lithuania, Poland, Russian Federation and Sweden.

reaction to concerns about the worsening state of the Baltic Sea marine environment. The unique and sensitive nature of the Baltic Sea (HELCOM 2009a), coupled with the fact that the sea is a valuable resource for the coastal countries, resulted in the Baltic coastal states acknowledging that the sustainability and well-being of the Baltic Sea depends on coordinated efforts and joint regional environmental standards. This initiative originated from the 1972 United Nations Conference on the Human Environment (UN 1972) which opted for regional cooperation in areas such as the Baltic where, for geographical and ecological reasons, regional cooperation could be carried out for a natural entity.

A revised Helsinki Convention was adopted in 1992 (HELCOM 1992) in order to take into account the further strengthening of environmental awareness which had paved the way for developments in international environmental law[2]; as well as in the region's political environment.[3] The 1992 Convention was signed by the environment ministers of the nine Baltic coastal states as well as by the Czech and Slovak Federal Republic, Norway and the European Economic Community and entered into force in 2000.

For three and a half decades HELCOM,[4] the governing body of the Helsinki Convention, has been working to protect the marine environment of the Baltic Sea. This work has been driven by the specific environmental, economic and social situation in the Baltic region and the special sensitivity of the Baltic Sea.

2.3 HELCOM: Over Three Decades of Work for the Baltic Sea

The Baltic Sea is the world's largest body of brackish water and ecologically unique. Due to its special geographical, climatological, and oceanographic characteristics, the Baltic Sea is highly sensitive to external impacts (Leppäranta and Myrberg 2009). Human activities, both in the sea itself and throughout its catchment area, have over the last centuries put considerable pressure on its marine ecosystem (see e.g. HELCOM 2002, 2009a, b).

---

[2] The 1992 Helsinki Convention included, amongst other things, the expansion of the 'Convention Area' to cover also inland waters within the catchment area of the Baltic Sea, rather than only the marine areas of the Baltic Sea. In addition, the revised Convention included new environmental principles such as the precautionary principle, polluter pays principle and the application of Best Environmental Practices (BEP) and Best Available Technology (BAT). The revised Conventions also took on board issues of biodiversity and nature conservation, as well as the sustainable use of the natural resources of the Baltic Sea area.

[3] With the fall of the Soviet Union, Estonia, Latvia and Lithuania became contracting parties to the Helsinki Convention (see also supra note 1).

[4] The Helsinki Commission, or HELCOM, works to protect the marine environment of the Baltic Sea from all sources of pollution through intergovernmental cooperation between the countries bordering the sea – Denmark, Estonia, Finland, Germany, Latvia, Lithuania, Poland, Russia and Sweden; as well as, the European Community.

2.3.1 The Work of HELCOM

In working towards the restoration and protection of the Baltic Sea marine environment, HELCOM has been using the best available scientific information to make informed regional policy recommendations. Since its establishment, HELCOM has cooperated closely with the scientific community in the coastal countries and elaborated joint guidelines for monitoring and reporting on the status of the Baltic Sea marine environment. This has made it possible for HELCOM to compile harmonized, region-wide environmental data which are the basis for regular thematic and holistic reports and assessments addressing the pressures to, and the status of, the Baltic marine environment. These assessments in turn make it possible to evaluate the effectiveness of existing management measures and to provide recommendations to decision makers on the need for further actions to reduce the impacts of human activities on the environment. In doing so, HELCOM acts as a bridge between science and policy.

In addition to implementing regional measures based on requirements imposed by other international organizations and agreements, HELCOM also supplements these by developing Recommendations[5] of its own according to the specific needs of the Baltic Sea. Furthermore, HELCOM works to ensure that commonly agreed upon environmental standards are fully implemented by all its contracting parties throughout the Baltic Sea and its catchment area.

In creating links between science and policy-making in the Baltic region, HELCOM has contributed to the development of measures that specifically take into account the unique environmental conditions and sensitivity of the Baltic Sea. The main priority areas of work carried out by HELCOM are based on the major environmental problems affecting the Baltic Sea, as defined by Baltic scientists and agreed upon by the contracting parties. At present, these are eutrophication, hazardous substances, biodiversity and nature conservation and maritime activities.

Despite that the focus of HELCOM activities is on regionally specific needs, much of the work carried out is linked to activities under other international forums, be these at a European or global level. Actions by HELCOM are carried out in three main ways: (1) Baltic Sea states may take joint initiatives to raise certain issues within international organizations – the International Maritime Organization (IMO) and the EU being good examples; (2) HELCOM also works towards harmonization and, where possible and needed, the strictest implementation of international environmental regulations in the region; and (3) HELCOM devises specific regional actions which either result in quicker actions than would be possible if acting on the global level or which are needed to ensure that regional interests are

---

[5] HELCOM Recommendations can be characterized as soft law instruments in that they are not legally binding as such. However, the fact that the Recommendations are adopted unanimously, and that countries are required to report on their national implementation, diminishes concerns about the lacking legal nature. Furthermore, the Recommendations support the implementation of the Helsinki Convention, which in turn is a legally binding instrument.

taken into account. This is done through development of relevant measures or joint initiatives, such as projects, seminars, awareness raising campaigns, etc.

2.3.2 Success Stories

Joint efforts by the HELCOM contracting parties have had some positive results, such as reductions in the pollution loads reaching the Baltic Sea and encouraging signs of improvements in the state of the environment. Examples include significant reductions in waterborne phosphorus loads, especially from point sources in the catchment area (Knuuttila 2009), such that almost all HELCOM countries have achieved the 50% reduction target as set by environment ministers in 1988 (HELCOM 1988). Emissions of heavy metals such as lead, cadmium and mercury to the air have dropped significantly, resulting in a reduction in atmospheric depositions of these metals on to the Baltic Sea of 71%, 44% and 26%, respectively (Gusev 2010a). Deposition of chlorinated dibenzo-*p*-dioxins/polychlorinated dibenzofurans (PCDD/Fs) on to the Baltic Sea has also decreased by 50% during the period 1990–2008 (Gusev 2010b).

Although the reductions in nutrient inputs have not reversed the serious eutrophication problem in the Baltic Sea, a positive observation is an increase in water transparency during the last 20 years detected in the Bornholm Sea and Arkona Sea (Fleming-Lehtinen and Kaartokallio 2009). The reductions in inputs of selected hazardous substances have resulted in decreasing concentrations of certain substances in some Baltic living organisms (Bignert et al. 2010a, b) and improvements have been observed in the status of populations of white-tailed eagle and seals (in the northern areas of the Baltic) (e.g. Helander et al. 2009; Bäcklin et al. 2009).

2.3.3 Further Needs

Regardless of the efforts already made and the successes achieved; anthropogenic pressures on the marine environment continue and there is still a clear need for further actions. Inputs of nutrients, especially from diffuse sources such as agriculture, are still excessive and the Baltic Sea continues to suffer from the effects of eutrophication (HELCOM 2009b). The extent of the loads and biological effects of persistent organic pollutants[6] and other hazardous substances are not well known and the entire Baltic Sea has a high contamination level (HELCOM 2010a).

[6] Persistent organic pollutants (POPs) bioconcentrate in organisms directly from the water via body surfaces, bioaccumulate from food, and biomagnify along the trophic chain from bacteria to predators.

In addition to eutrophication and hazardous substances, physical alteration of habitats, fishing, alien species, maritime traffic, climate warming and other anthropogenic pressures are threatening the biodiversity of the Baltic Sea (HELCOM 2009a).

Even though necessary further measures have been identified by HELCOM, work is challenged by various factors. Firstly, it can take a long time before the beneficial effects of selected measures are seen in the environment. Secondly, in the face of competitive interests between different sectors and limited resources, environmental needs often do not receive the resources and attention they deserve and need. Thirdly, the enlargement of the EU has meant that most HELCOM contracting parties are prioritizing EU legislation and that many decision-making tasks have shifted from national level to EU level, such as in the case of the EU Common Agricultural Policy[7] and the EU Common Fisheries Policy.[8] Despite this, HELCOM still sees its role in the agricultural and fisheries fields as very important, especially as HELCOM holds information on the status of the marine environment, which should be the basis on which management decisions are made.

2.4 The HELCOM Baltic Sea Action Plan (BSAP): A New Environmental Strategy

In November 2007, the environment ministers of the HELCOM countries adopted an ambitious strategy, the HELCOM BSAP, which aims to reduce pollution to the Baltic Sea and reverse its degradation by 2021 (HELCOM 2007a). The BSAP is the first attempt by a regional seas convention to incorporate the ecosystem-based approach to the management of human activities[9] into the protection of the marine environment. The novelty of the approach used in the HELCOM action plan is that it puts the ecosystem at the centre, by defining the status of the sea as we want it to be in the future, and focusing management decisions on this goal as opposed to the more traditional approach of addressing the sources of pollution on a sector by sector basis, without directly linking the measures to the status of the Baltic Sea.[10]

[7] For more information about the EU Common Agricultural Policy (CAP), see: http://europa.eu/pol/agr/index_en.htm. Accessed 24.02.2011.

[8] For more information about the EU Common Fisheries Policy (CFP), see http://ec.europa.eu/fisheries/cfp/index_en.htm. Accessed 24.02.2011.

[9] The ecosystem approach to the management of human activities was officially accepted in the 1992 Rio Declaration on Environment and Development (UN 1992) and reiterated in the Plan of Implementation of the 2002 World Summit on Sustainable Development (UN 2002).

[10] Examples of the historical HELCOM approach are the 1988 and 1998 HELCOM Ministerial decisions on a flat 50% reduction target for discharges of nutrients and hazardous substances (HELCOM 1988, 1998).

The holistic plan contains concrete actions to solve the major problems affecting the Baltic Sea, including measures to curb eutrophication; prevent pollution involving hazardous substances; improve maritime safety and accident response capacity; and halt habitat destruction and the decline in biodiversity. The fourth priority, biodiversity, is especially interlinked to the other issues, because without improvement in curbing inputs of pollutants, the state of biodiversity will not improve. In fact the status of Baltic Sea biodiversity can also be viewed as a measure of the progress made towards successfully implementing the action plan.

2.4.1 Cornerstones of the BSAP

The BSAP aims to be an overarching instrument which has as its *vision* and main target the achievement, by 2021, of a Baltic Sea in good ecological status – a sea with diverse biological components functioning in balance and supporting a wide range of sustainable economic and social activities. This vision is supported by four main strategic *goals* addressing issues of particular concern for the Baltic marine environment, namely eutrophication, hazardous substances, maritime activities and biodiversity.

Each of the main goals of the action plan is further defined by *ecological objectives* which describe the characteristics of the sea which we aspire towards. These ecological objectives define good ecological status and aim to be easily understandable. Example ecological objectives include clear water, an end to excessive algal blooms, and viable populations of species.

The complete framework of vision, goals and ecological objectives used in the BSAP is demonstrated in Fig. 2.1. As the action plan follows adaptive management principles, the objectives and targets will be periodically reviewed and revised using a harmonized approach and latest available scientific information.

2.4.2 Assessing the Progress Toward the BSAP Objectives

In order to determine whether management measures for protecting the Baltic Sea are achieving our commonly agreed goals for the marine environment, the ecological objectives need to be made operational. For each ecological objective, indicators, based on easily and accurately measureable parameters, need to be developed. It is important that the data are comparable over time and between countries in order to ensure Baltic-wide, harmonized datasets.

Reference values for the different indicators are defined by an understanding of what "pristine conditions" have been and these are determined either by historic data or by modeling. According to the EU Water Framework Directive (WFD, EC 2000), the reference condition for a given indicator is a description of the chemical and biological quality elements that exist, or would exist, at high status, that is, with no or only very minor disturbance from human activities.

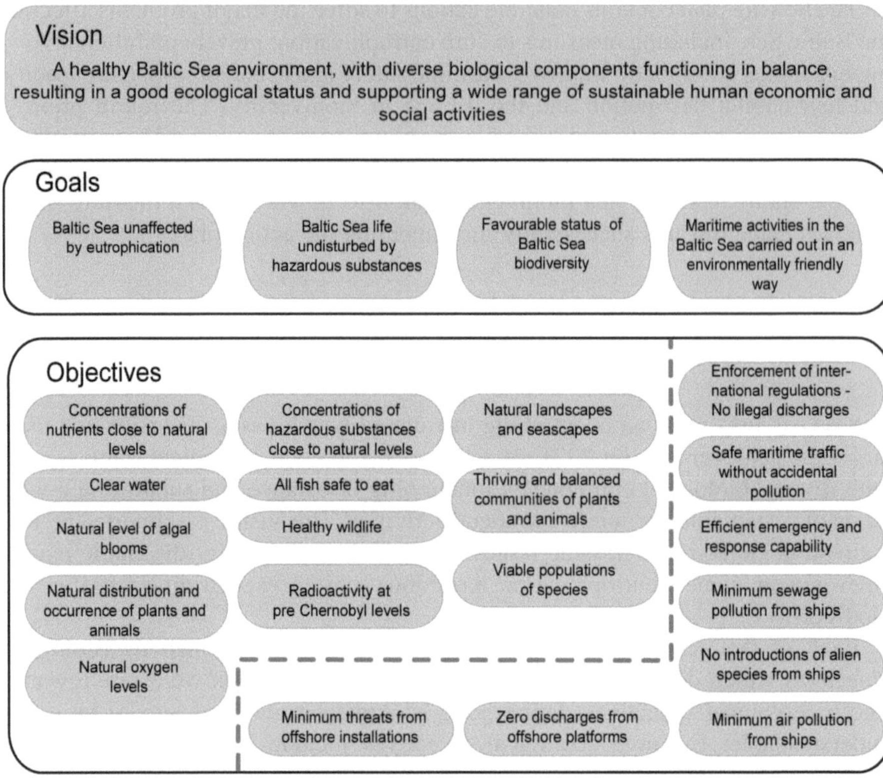

Fig. 2.1 Vision, goals and ecological objectives of the HELCOM Baltic Sea Action Plan

An *acceptable deviation* from the reference condition (usually defined by expert judgment) is used to calculate specific *target values* for each indicator. An acceptable deviation is considered to be synonymous with the border between good and moderate ecological status (for more information, see Andersen et al. 2010 and HELCOM 2006). It should be noted that because environmental conditions naturally differ greatly in different areas around the Baltic Sea, the reference values, acceptable deviations and target values also need to be adjusted accordingly.

The progress towards achieving the agreed ecological objectives can be assessed by calculating the Ecological Quality Ratio (EQR) which represents the relationship between the observed value (current status) of an ecological parameter and the value for that parameter in reference conditions.[11] The EQR is then used to determine the ecological status of an area as shown in Fig. 2.2.

[11] The EQR approach has been used in the HELCOM integrated thematic assessments on biodiversity and eutrophication (HELCOM 2009a, b). The HELCOM integrated thematic assessment on hazardous substance (HELCOM 2010a), however, is based on a different approach. For most persistent organic pollutants the reference concentration is zero, or close to zero, and therefore it

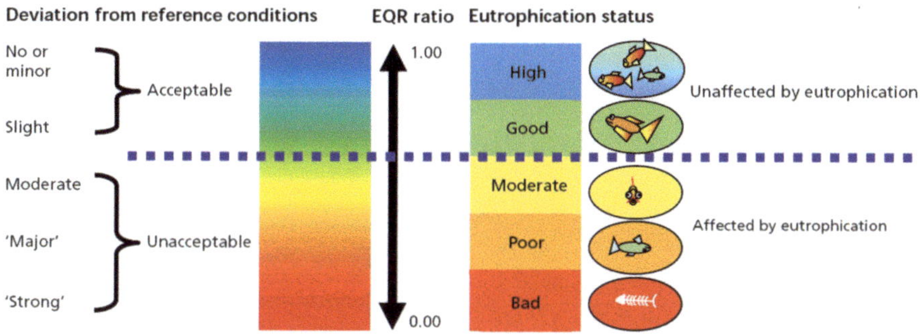

Fig. 2.2 Overview of the EQR concept and its use for classifying water bodies affected by eutrophication (Source: HELCOM 2009b)

An example of regional variations in reference conditions, acceptable deviations and targets is illustrated in Fig. 2.3 where these values are given for water transparency (Secchi depth) in different sub-basins of the Baltic Sea.

In May 2010, the ministers of the environment of the HELCOM parties adopted a demonstration set of core eutrophication indicators which allow for assessing the progress towards reaching the ecological objectives defined in the BSAP. So far, core indicators have been developed for concentrations of nitrogen and phosphorus, water clarity, chlorophyll-*a* concentrations and the status of benthic invertebrate communities.[12] These have further been combined to produce an integrated indicator describing the eutrophication status. For more information about indicators and the methodologies used to calculate targets, status etc., see HELCOM (2009b, 2010b) and Andersen et al. (2010).

2.4.3 *Actions for Achieving a Healthy Baltic Sea*

The development of the actions and measures agreed upon in the BSAP was coordinated by a task force group including all HELCOM members, as well as representatives from nongovernmental and governmental organisations. International financial institutions, e.g. Nordic Investment Bank (NIB) and Nordic Environmental Finance Corporation (NEFCO), also participated in the work by giving input on how to ensure future funding of concrete projects aimed at implementing

makes sense to instead of the EQR, use a threshold value as the target (the threshold level being the level above which the substance is known to cause adverse effects). By comparing the measured value (status) with the threshold value we get a Contamination Ratio (CR). If status is above the threshold, then CR is above one and we move from an acceptable status to an unacceptable status.

[12] The core set of eutrophication indicators can be viewed on the HELCOM website http://www.helcom.fi/BSAP_assessment/en_GB/main/. Accessed 24.02.2011.

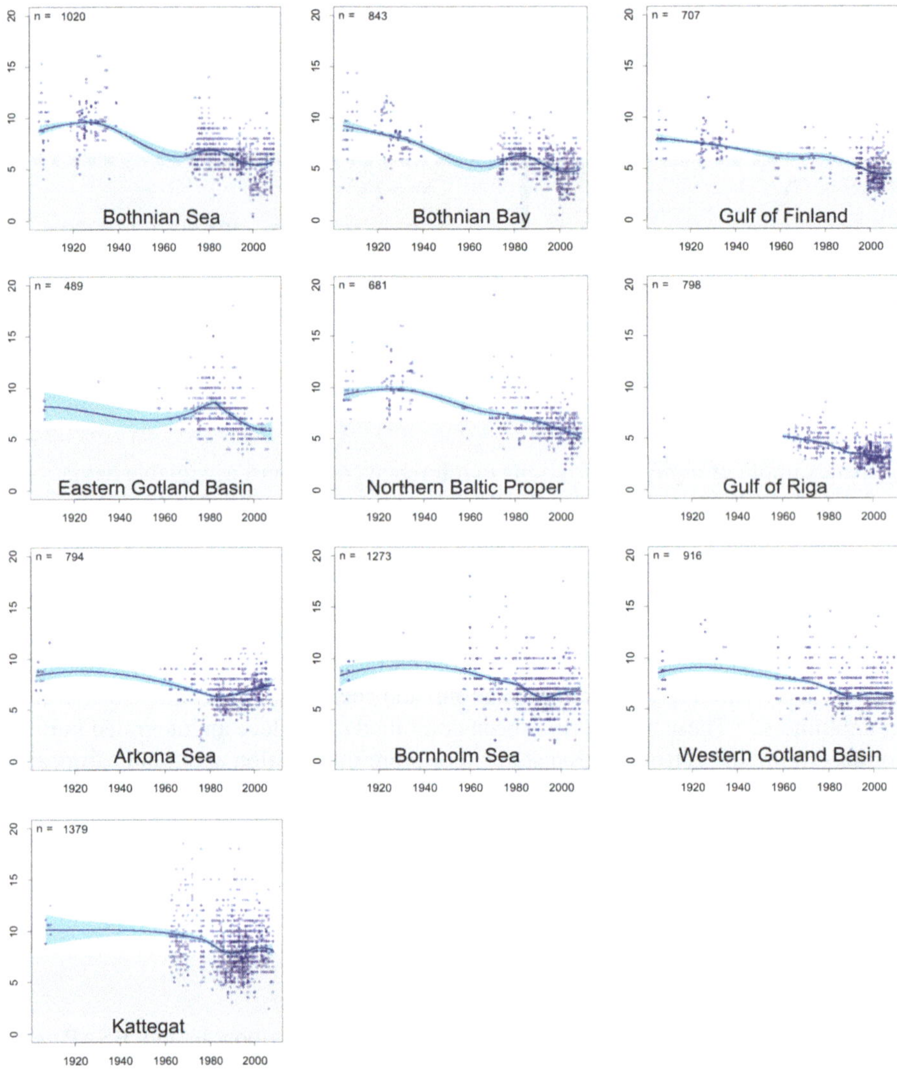

Fig. 2.3 Water transparency in June-September measured as Secchi depth (m) between years 1903 and 2009 in the open-sea sub-regions of the Baltic Sea. Secchi depth observations (m) are plotted against the year of observation and the curves fitted with non-linear smoothing (blue line) and shown with 95% confidence intervals (light blue area). The level below the target value set by the HELCOM thematic assessment on eutrophication (HELCOM 2009b) is coloured light gray. The number of observations (n) is shown on each figure. (Source: Fleming-Lehtinen and Laamanen, in press)

the action plan (e.g. improving wastewater treatment). The final phases of the development comprised political discussions between coastal countries and the European Commission, with other stakeholders participating as observers. Parallel HELCOM assessment activities (e.g. HELCOM 2006) provided the necessary scientific consensus for exact targets to be reached by the actions (Backer et al. 2010).

The BSAP actions include deadlines for achieving the agreed targets for each of the main environmental priorities. In addition to defining common actions to be carried out by all HELCOM contracting parties, the BSAP also requires the countries to develop their own tailor-made national implementation programmes which outline the most cost-effective measures for achieving the environmental targets in each country. The action plan also distinguishes between measures that can be implemented at regional or national level, and measures that can only be implemented at EU level (e.g. Common Fisheries Policy, Common Agricultural Policy, and controls over the marketing and use of chemicals) or globally (e.g. the shipping controls defined by the IMO).

One of the major highlights of the action plan is that it opens a new era in marine environmental protection by including the concept of maximum allowable nutrient input, which still makes it possible for the Baltic Sea to reach a good ecological status (see Table 2.1). The Baltic Sea coastal countries acknowledged in the plan that 'there is a need to reduce the nutrient inputs and that the needed reductions shall be fairly shared by all Baltic Sea countries'. Initial estimates of the quantities of nutrient reductions needed to reach the environmental targets as concerns eutrophication were calculated using a model developed by the Swedish MARE programme (Wulff et al. 2007). In addition, scenarios were elaborated on how far the full implementation of existing HELCOM Recommendations, as well as EU legislation and programmes, would bring the Baltic Sea towards the agreed ecological objectives for eutrophication, using the target *clear water* as a basis. These results

Table 2.1 The maximum allowable nutrient inputs (in tonnes per year) to each Baltic Sea sub-region if good environmental status is to be reached and the corresponding necessary nutrient reductions as agreed upon in the HELCOM Baltic Sea Action Plan

Sub-region	Maximum allowable input		Inputs in 1997–2003		Needed reductions	
	P	N	P	N	P	N
Bothnian bay	2,580	51,440	2,580	51,440	0	0
Bothnian sea	2,460	56,790	2,460	56,790	0	0
Gulf of Finland	4,860	106,680	6,860	112,680	2,000	6,000
Baltic proper	6,750	233,250	19,250	327,260	12,500	94,000
Gulf of Riga	1,430	78,400	2,180	78,400	750	0
Danish straits	1,410	30,890	1,410	45,890	0	15,000
Kattegat	1,570	44,260	1,570	64,260	0	20,000
Total	21,060	601,720	36,310	736,720	15,250	135,000

Note: *P* phosphorus, *N* nitrogen (units: tonnes)
Source: HELCOM (2007a)

Table 2.2 Country-wise provisional nutrient reduction requirements[a] as agreed upon in the HELCOM Baltic Sea Action Plan

Country	Phosphorus (t)	Nitrogen (t)
Denmark	16	17,210
Estonia	220	900
Finland	150	1,200
Germany	240	5,620
Latvia	300	2,560
Lithuania	880	11,750
Poland	8,760	62,400
Russia	2,500	6,970
Sweden	290	20,780
Transboundary common pool[b]	1,660	3,780

[a]This refers to the reduction requirement from the average annual pollution load from the country during the period 1997–2003
[b]In the BSAP, the HELCOM contracting parties agreed that "transboundary pollution originating in the non-contracting states Belarus and Ukraine should be addressed by initiating joint activities e.g. by bi- and/or multilateral projects and through other existing funding mechanisms as well as by international agreements such as the 1992 UNECE Convention on Transboundary Waters and Lakes, and the River Basin Management Plans of the EU Water Framework Directive for HELCOM Contracting States being also EU Member States
Source: HELCOM (2007a)

were then used to calculate specific nutrient reduction targets and to identify actions related to reducing nutrient loading to the Baltic Sea.[13]

The MARE model calculated that of the average annual loads to the whole Baltic Sea, reductions of an estimated 15,250 t of phosphorus and 135,000 t of nitrogen would be necessary to achieve agreed targets. Similarly, quantitative reduction requirements were calculated for each of the sub-basins (see Table 2.1) and provisional allocations of nutrient reduction requirements were given to each HELCOM country and to transboundary loads (see Table 2.2). These figures are all contained in the BSAP, with the idea of allowing the contracting states more flexibility in choosing the management actions that are nationally most suitable and cost-effective to reach the agreed targets.

At the time of the adoption of the BSAP, it was recognized that the calculated maximum allowable nutrient loads and the country-wise allocations were based on the best knowledge available at the time and that reviewing and revising of the figures would be necessary as soon as more updated data, information and more advanced models would be available. At the time of writing this chapter, HELCOM is working on further developing the environmental targets for eutrophication

[13] In the BSAP, the HELCOM contracting parties noted that the calculated figures were based on the best available scientific information, and thus stressed the provisional character of the data and that all the figures related to targets and maximum allowable nutrient inputs should be periodically reviewed and revised using a harmonized approach using updated information.

indicators as well as refining the procedure for assessing annual nutrient loads and thus the progress of the HELCOM countries towards reaching their provisional nutrient reduction allocations. This work will provide input to the reviewing, and if necessary revision, of the maximum allowable nutrient inputs agreed upon in the BSAP as well as the country-wise nutrient reduction allocation schemes which are necessary for achieving the environmental targets, which in turn may result in the need to identify further measures and actions. This continuous assessment and revision process is one example of the application of adaptive management under the BSAP.

2.4.4 Implementation and Financing of the BSAP

Since the adoption of the BSAP, the HELCOM countries have been working towards implementation of the actions agreed upon in the plan. One of the first tasks for the HELCOM countries has been to prepare national implementation programmes which were to be presented at the HELCOM Moscow Ministerial Meeting in May 2010 (HELCOM 2010c).[14] In these national implementation programmes, the countries should define what measures they are applying (or planning to apply) at a national level, in order to fulfill the objectives of the BSAP. Furthermore, countries should present what magnitude of reductions in nutrient inputs are expected to follow the implemented measures as well as the cost-effectiveness of the different measures.

Due to the transboundary nature of pollution, the BSAP also addresses pollution loads to the Baltic Sea that originate from non-HELCOM countries. A significant portion of waterborne pollution loads originate from non-coastal states within the Baltic Sea drainage area and to address these sources, HELCOM is working to support the establishment of bi- and multi-lateral projects, involving also private initiatives funded by private foundations and international financial institutions.

Distant sources and Baltic Sea shipping contribute about 40% of airborne nitrogen deposited onto the Baltic Sea, thus significantly adding to the existing eutrophication problem (HELCOM 2009b). Because emissions from shipping and some non-HELCOM European countries are the source of a notable amount of the nitrogen input to the Baltic Sea, HELCOM is working to influence emission policies under IMO, the EU National Emission Ceiling Directive (NECD, EC 2001) and the Gothenburg Protocol of the United Nations Economic Commission for Europe Convention of Long-Range Transboundary Air Pollution (UNECE 1999).

[14] Information about the status of implementation of the BSAP by the HELCOM countries can be viewed on the HELCOM website: http://www.helcom.fi/BSAP/Implementation/en_GB/Implementation/. Accessed 24.02.2011.

The main sources of funding for implementing the BSAP actions include national budgets and EU structural funds, including the EU Cohesion Fund which aims to help new member states to implement EU Directives.[15] Russia, as a non-EU HELCOM contracting party, will benefit from funding provided for high priority environmental projects through e.g. bilateral agreements and the Northern Dimension Environmental Partnership fund.[16] In 2009, in an effort to speed up implementation of the BSAP actions, NIB and NEFCO set up a BSAP Fund that aims to provide grants for technical assistance to projects that support the implementation of the BSAP.[17]

Despite all the identified objectives, measures and actions in the BSAP, there is still much work to be done to further elaborate on them, especially as concerns issues which are not so well understood yet. In comparison to eutrophication, much less is known about the loads and the biological effects of different hazardous substance. Also there is a large knowledge gap as concerns Baltic Sea biodiversity, as a great deal still needs to be learned about the distribution and status of species, communities and habitats. Information which is essential for developing operational indicators of environmental status and without a good understanding of all the pressures and their effects on the ecosystem, it is not possible to define clear and efficient management objectives.

In an effort to reduce information and management gaps, HELCOM is presently working on developing core indicators related to hazardous substances and biodiversity. The main aim of these quantitative indicators is to allow for the assessing of the progress (and effectiveness of measures) towards reaching the BSAP ecological objectives, strategic goals and overall vision. In the progress of developing the core set indicators, HELCOM will review existing monitoring programmes and propose revisions to the programmes so that national monitoring activities will produce the data necessary for regular assessments.

To solve the environmental problems of the Baltic Sea, commitment by the HELCOM contracting parties to the implementation of the BSAP is of utmost importance. There are numerous challenges which complicate the implementation of the action plan and much remains to be done to further develop actions addressing the problems of hazardous substances and threats to biodiversity. Cooperation needs to be increased between different sectors (such as between the environmental, agricultural, fisheries and maritime authorities as well as between public and private sectors) in order to find joint solutions to the pressures affecting the Baltic Sea. Applying the concept of sustainable use of resources is not easy when economic interests often compromise the well-being of the environment.

---

[15] For more information on the EU Cohesion Fund, see: http://ec.europa.eu/regional_policy/funds/procf/cf_en.htm. Accessed 24.02.2011.

[16] For more information about the Northern Dimension Environmental Partnership fund, see: http://www.ndep.org/home.asp?type=nh&pageid=6#support. Accessed 24.02.2011.

[17] For more information, see: http://www.nefco.org/files/48521_BSAPfactsheet.pdf. Accessed 24.02.2011.

And it is no surprise that, especially during financially challenging times, other concerns are likely to be prioritized over environmental protection activities. On a positive note, however, many governments and private organizations are beginning to see the economic benefits of protecting the environment, not only by recognizing the value of ecosystem services, but also by applying the concept of green growth and investing in the development of green jobs and technology.

Adaptive management needs to be employed by ensuring that appropriate measures are identified and implemented and new and improved information becomes available. This is not always easy, however, in the face of scientific uncertainty and the cost and effort of transposing new measures into national laws.

For the BSAP to be successful, strong political will, financial and human resources, as well as involvement by all relevant stakeholders is necessary. Also, as many activities affecting the Baltic Sea marine environment are managed from outside the Baltic Sea region, it is very important that regional environmental needs are taken into account when developing measures within other international fora. This should be done by further enhancing cooperation with relevant international institutions and other regional agreements.[18]

2.4.5 Links Between the BSAP and Other Multilateral Environmental Agreements

HELCOM generally complements and contributes to the harmonized implementation of various international legislative frameworks. This is also the case with the BSAP; as the implementation of the plan by HELCOM contracting parties is both a fulfillment of the obligations under the Helsinki Convention and an essential contribution to the fulfillment of the obligations of the contracting parties under various other international legal frameworks.

In developing the BSAP, HELCOM took into account the environmental provisions of the Maritime Doctrine of the Russian Federation. Close co-operation with Russia, which is the only non-EU member country in HELCOM, is crucial for further progress to be made in rescuing the Baltic marine environment. The action plan is also instrumental for the implementation of the renewed Northern Dimension policy,[19] the Baltic Sea regional aspects of the EU-Russian

[18] Examples are the management of shipping activities under the International Maritime Organization, the management of fisheries and agriculture under the EU framework and the addressing of transboundary air pollution under the UNECE Convention on Long Range Transboundary Air Pollution.

[19] For more information, see: http://ec.europa.eu/external_relations/north_dim/index_en.htm. Accessed 24.02.2011.

Environmental Dialogue,[20] the Nordic Environmental Action Plan,[21] and EU policies such as the European Maritime Policy,[22] the WFD and the Marine Strategy Framework Directive (MSFD, EC 2008).

The BSAP has been widely supported by politicians at various forums and the European Union has described the plan as a cornerstone for further action in the Baltic Sea region, taking on board much of the BSAP in the environment and security pillars of the recently adopted EU Strategy for the Baltic Sea (EC 2009a). The MSFD foresees an action plan for each eco-region, including the Baltic Sea; and the BSAP is currently being considered as a possible pilot project for the regional implementation of the directive. HELCOM is in a unique position to both deliver this, given its embracing of all the countries in the Baltic Sea catchment area, and also to ensure that the special characteristics of the Baltic Sea are fully accounted for in European policies.

At the 2010 HELCOM Moscow Ministerial Meeting, the Ministers of Environment of the HELCOM countries decided "to establish, for those HELCOM Contracting States being also EU Member States, the role of HELCOM as the coordinating platform for regional implementation of the EU MSFD in the Baltic Sea including striving for harmonized national marine strategies for achieving good environmental status according to the HELCOM Baltic Sea Action Plan and the EU MSFD." This decision is an affirmation that different international agreements need to be streamlined in order to avoid duplication of work and is in line with Articles 5[23] and 6[24] of the EU MSFD. Examples of regional coordination include the ongoing HELCOM work to develop core indicators and to revise HELCOM monitoring programmes in such a way that they also fulfil the requirements under the EU WFD[25] and MSFD.[26]

[20] For more information, see: http://ec.europa.eu/environment/enlarg/russianis_en.htm. Accessed 24.02.2011.

[21] For more information, see: Nordic Council of Ministers' Nordic Environment Action Plan 2009–2012. http://www.norden.org/en/publications/publications/2008-733. Accessed 24.02.2011.

[22] For more information about the Integrated Maritime Policy for the European Union, see: http://ec.europa.eu/maritimeaffairs/subpage_en.html. Accessed 24.02.2011.

[23] Article 5§2 of the MSFD states that: Member States sharing a marine region or subregion shall cooperate to ensure that, within each marine region or subregion, the measures required to achieve the objectives of this Directive [...] are coherent and coordinated across the marine region or subregion concerned.

[24] Article 6§1 of the MSFD states that: In order to achieve the coordination referred to in Article 5§ 2, Member States shall, where practical and appropriate, use existing regional institutional cooperation structures, including those under Regional Sea Conventions, covering that marine region or subregion.

[25] Especially Article 5 of the WFD on the determination of the characteristics of the river basin district, review of the environmental impact of human activity and economic analysis of water use, as well as, Article 8 of the WFD on the establishment of programmes for the monitoring of water status.

[26] Specifically Articles 5§2(a) concerning (1) the preparation of an initial assessment, (2) a determination of good environmental status for the waters concerned, (3) establishment of a series

As a front-runner in the application of the ecosystem approach, the HELCOM action plan could also serve as a model example to be followed by the Regional Seas Conventions and Action Plans under the auspices of the United Nations Environmental Programme's Regional Seas Programme.[27]

2.5 Discussion and Conclusions

2.5.1 Importance of Improving the Link Between Science and Policy

Science is the foundation of the work of HELCOM and has also had a central role in the elaboration of the BSAP. Interaction between science and policy is central not only to developing new policies, but more importantly, should support the application of adaptive management. Adaptive management is an iterative process of refining management practices over time as new information becomes available and is central also to the implementation of the ecosystem approach (Mee 2005).

An example of the interface between science and policy in the HELCOM framework, in particular as concerns tackling the serious eutrophication problem being faced by the Baltic Sea, is illustrated in Fig. 2.4. The ongoing revision process for updating the provisionally adopted *maximum allowable nutrient input* values and *country-wise nutrient reduction targets* which were adopted with the BSAP in 2007 is a good example of how adaptive management is being put into practice under the HELCOM umbrella. These figures are being reviewed and corrected using more recent data on the magnitude and sources of pollution loads, better defined environmental targets as well as improved models. A similar application of adaptive management is expected in the future for also other components of the action plan (e.g. biodiversity and hazardous substance) after evaluation of whether agreed measures are sufficient to reach good environmental status in the Baltic Sea.

Without sound scientific knowledge, informed policy making would not be possible. Monitoring activities allow scientists to examine status and trends and to assess the effectiveness of actions and measures taken – information which in turn is used to elaborate strategic goals, management options and policy recommendations.

---

of environmental targets and associated indicators and (4) establishment and implementation of a monitoring programme for ongoing assessment and regular updating of targets and Article 8 on the specifics of the initial assessment of national marine waters.

[27] The UNEP Regional Seas Programme was launched in 1974 in the wake of the 1972 United Nations Conference on the Human Environment held in Stockholm (UN 1972). For more information, see http://www.unep.org/regionalseas/about/default.asp. Accessed 08.01.2010.

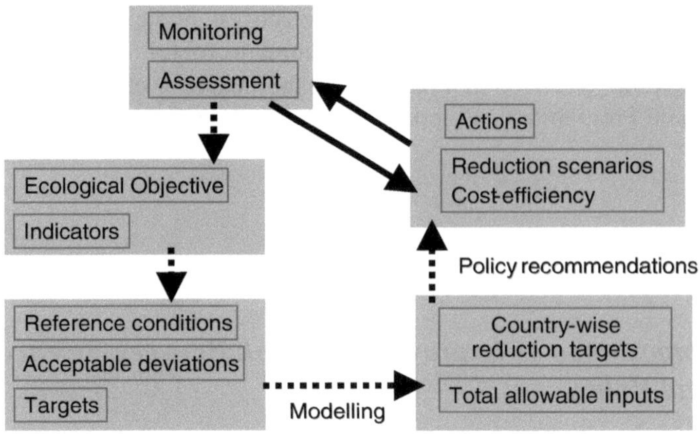

Fig. 2.4 Eutrophication management cycle of the Baltic Sea Action Plan

A sound scientific basis for such assessments requires harmonized monitoring methods as well as long-term data series which allow for integration of comparable datasets.

Fortunately, there has been joint regional monitoring of the Baltic Sea for several decades (see for example the HELCOM COMBINE Manual[28]) meaning that there exists a great deal of comparable data for the region. Nevertheless, there are gaps in data, especially regarding historic data that is necessary for defining reference conditions, and subsequently environmental targets, for indicators describing the status of the Baltic Sea. Also, not all parameters have been (or are being) monitored, e.g. in the case of many chemicals which are new and occur in very small, but potentially highly toxic, doses for which toxicological responses, especially when combined with other substances, are not well understood.

As already mentioned earlier in this chapter, HELCOM is presently working on further developing core indicators and reviewing the current monitoring activities in the region. An outcome of this work will be proposals for revising the monitoring activities in the region so that they are cost efficient and support reporting and assessment requirements for HELCOM, EU directives as well as other international agreements. This work, which involves both scientists and decision-makers, can be seen as a move towards further improving the links between science and policy.

[28] The Manual for Marine Monitoring in the COMBINE Programme of HELCOM is available on the HELCOM website at: http://www.helcom.fi/groups/monas/CombineManual/en_GB/main/. Accessed 24.02.2011.

2.5.2 *Linking Climate Change to Baltic Sea Protection*

2.5.2.1 The Effects of Climate Change on the Baltic Marine Environment

Climate related ecosystem changes are already taking place in the Baltic Sea region.[29] Recent regional climate projections suggest that near-surface air temperatures will further increase by 3–5°C in the Baltic Sea area during this century and that precipitation in the northern Baltic Sea drainage basin will increase (Graham et al. 2008). Shorter and wetter winters are likely to also result in decreased snow and ice cover, leading to greater run-off from the drainage area and increased inflow of water (and nutrients) to the sea. Expected consequences in the marine environment include decreased salinity as well as increases in sea surface temperature,[30] sea level, and acidification of sea water (Belkin 2009; Graham et al. 2008; Sabine et al. 2004).

The projected changes in climate as a consequence of global warming may have profound implications for Baltic biodiversity (Dippner et al. 2008; Granskog et al. 2006; HELCOM 2009a; Laine 2008; Nielsen and Kvaavik 2007). In particular, the predicted decrease in salinity, as a result of increased precipitation, is expected to shift the distribution limits of several important habitat-structuring species and key species in the Baltic ecosystem, such as bladder wrack, eelgrass, blue mussel and cod. The projected increase in temperature and decrease in ice cover is also likely to impact the ranges of species. For example, decreasing ice cover is likely to directly impact ringed seals that breed on ice and, with increasing temperatures, alien species of southern origins may establish viable populations in the Baltic Sea.

The effects of global climate change are not only expected to result in additional pressures to the Baltic ecosystem, but also to exacerbate the effects of other anthropogenic pressures on the environment (HELCOM 2007b). For instance, the recently published HELCOM thematic assessment on biodiversity points out that climate-driven changes in the marine ecosystem (including changes in salinity and sea surface temperature, as well as deep-bottom oxygen depletion) have enhanced the negative impacts of eutrophication and fisheries on the marine environment (HELCOM 2009a; Sherman et al. 2009).

There are also likely to be other indirect impacts to the marine environment caused not only by climate change but also by policies designed to combat it. The HELCOM biodiversity assessment suggests that actions, such as the requirements adopted by the EU to, by 2020, cut CO_2 emissions by 20% from 1990 levels and to achieve the level of 20% of renewable energy of all consumed energy in the EU

---

[29] It is not the aim of this chapter to go into details about the effects of climate change on the Baltic Sea and therefore the reader is encouraged to explore the other relevant chapters of this book for more specific information.

[30] According to Belkin (2009) compared to other large marine ecosystems around the world, the Baltic Sea has shown the highest rate of increase in water temperature during the past half a century.

(EC 2009b), may have indirect negative effects on Baltic Sea biodiversity. It is highly likely that such a target will result in an increased number of offshore wind farms in the Baltic Sea, putting further pressure on the use of the marine space in some areas. Growing demand for carbon capture and sequestration technologies and sites, being proposed by EU institutions, may mean that potential seafloor sites will be explored with yet unknown effects on benthic ecosystems. The energy targets will also likely be linked to installations of new underwater cables and pipelines. An increase in bioenergy production, such as increased cultivation of energy-producing crops, may result in an increase in use of currently set aside land and also of chemical fertilizers and thus contribute to increased eutrophication.

2.5.2.2 The Impacts of Climate Warming on Baltic Sea Protection Measures

The effects of climate warming will no doubt require addition management and protection measures if we are to achieve our agreed goals and objectives for a healthy Baltic Sea. The BSAP acknowledges the significance of climate change and the possible need for more stringent measures to protect the Baltic Sea in the future[31] but the specific measures and actions currently identified in the plan do not take into account the impacts of climate change as such.

Tackling the eutrophication problem of the Baltic Sea will be especially challenging if the predicted climate warming and increased precipitation become a reality as increased agricultural activities (resulting from e.g. an extended growing season and greater demands for food and energy crops) and run-off are likely to lead to an increase in nutrient loading to the sea. In addressing eutrophication, the Baltic Sea countries have already taken numerous measures to reduce the loading of nutrients to the sea, particularly from point sources. This means that the proportion of nutrient inputs originating from diffuse sources such as agriculture has increased (Knuuttila 2009). If nutrient inputs increase, then the present measures will not suffice for reaching the objectives and goals of the action plan.

The necessary adjustment of management measures (adaptive management) will require effective and continuous feedback between different activities such as monitoring programmes and management measures, and importantly, the results and recommendations of assessments and analyses need to be turned into appropriate decisions and properly implemented.

[31] The preamble of the BSAP states: **FULLY AWARE** that climate change will have a significant impact on the Baltic Sea ecosystem requiring even more stringent actions in the future and of the efforts made by the Conference of the Parties to the 1992 United Nations Framework Convention on Climate Change.

2.5.2.3 How Can HELCOM Work Help in Adaptation to the Effects of Climate Change?

Climate change and protection of the Baltic Sea marine environment go hand in hand. On the one hand, climate change will challenge the work to restore the Baltic marine environment, and on the other hand, protecting the marine ecosystems may help to reduce the impacts and facilitate adaptation to the changes resulting from climate change. Climate change therefore increases the importance of working towards also reducing other human induced pressures on the Baltic Sea marine environment. Box 1 outlines some examples of why HELCOM's work is important in the face of climate change.

The Helsinki Convention (the implementation of which the BSAP is facilitating) has as one of its fundamental principles the *precautionary principle*[32] which requires that action should be taken at source if an adverse effect on the environment is likely, even if a clear cause-and-effect relationship cannot be established.[33] Because the effects of measures may take decades before becoming apparent in the marine environment – due, for example, to the slow exchange rate of water in the Baltic Sea – the importance of taking early action cannot be over-emphasized. It is therefore crucial to identify ways of minimizing the gap between science and policy, especially as concerns making decisions based on uncertain information.

On a more positive note, despite it seeming that the incorporation of climate change into HELCOM work is still in its infancy, in actual fact one could argue that much of the work already carried out under the umbrella of HELCOM supports the mitigation of the effects of climate change. In other words, by working towards achieving a healthy marine environment, we are also working towards having an ecosystem that is more resilient to different pressures and better able to adapt to climatic changes.

Why Is HELCOM's Work Important in the Face of Climate Change?

- Biodiversity provides insurance against external pressures. The Baltic Sea is species poor and each species has an important role in buffering the ecosystem against extreme and potentially long-lasting changes. Reducing eutrophication, diminishing harmful impacts from fisheries and hazardous substances, eliminating pollution from maritime activities as well as establishing an ecologically coherent network of protected areas will help to safeguard biodiversity also in a changing climate.
- Achieving and maintaining the natural functions of the Baltic Sea ecosystem is crucial. Biogeochemical cycling of nutrients has already been altered by humans as the overdose of nutrients from external (e.g. land-

(continued)

[32] Article 3§2 of the 1992 Helsinki Convention.

[33] For an in depth discussion about the application of the precautionary principle in the work of HELCOM, see Pyhälä et al. (2007).

based and atmospheric) sources has contributed to extensive anoxia and dead zones at the bottom of the sea. Climate change threatens to aggravate the situation, hence it is even more crucial to decrease nutrient loads to the sea.
- Hazardous substances cause extra stress to organisms which are already living close to their low-salinity tolerance limit in the Baltic Sea. Climate change induced higher temperatures and further decreased salinity may affect the ability of these species to cope with toxic substances. Reduction of the inputs of hazardous substances to the Baltic Sea is essential for improving the resilience of organisms to the effects of climate change.
- The suite of actions and measures outlined in the BSAP has been tailored to achieve a healthy Baltic Sea in terms of biodiversity, eutrophication and hazardous substances, as well as environmentally sound maritime activities. Ultimately such a sea is also better prepared for the extra challenges caused by climate change.

2.6 Food for Thought

During the BALTEX summer school course *Climate impacts on the Baltic Sea: from science to policy*, one of the student tasks was to identify whether and how the objectives of the BSAP might be affected by expected climate change, and to suggest ways that the action plan could be modified to accommodate these impacts.

When considering how to better integrate the issue of climate change with the BSAP, some questions to take into account might be:

- What are the goals and objectives for responding to the threat of climate change in the Baltic Sea region?
- What parameters could be used as measureable indicators?
- What stakeholders should be involved; also from outside the Baltic Sea region?
- How to enhance links between science and policy?
- Which are the key policy instruments and frameworks that need to be integrated to avoid duplication of work?

References

Andersen JH, Axe P, Backer H, Carstensen J, Claussen U, Fleming-Lehtinen V, Järvinen M, Kaartokallio H, Knuuttila S, Korpinen S, Kubiliute A, Laamanen M, Lysiak-Pastuszak E, Martin G, Murray C, Møhlenberg F, Nausch G, Norkko A, Villnäs A (2010) Getting the measure of eutrophication in the Baltic Sea: towards improved assessment principles and methods. Biogeochemistry. doi:10.1007/s10533-010-9508-4

Backer H, Leppänen J-M, Brusendorff AC, Forsius F, Stankiewicz M, Mehtonen J, Pyhälä M, Laamanen M, Paulomäki H, Vlasov N, Haaranen T (2010) HELCOM Baltic Sea Action Plan – a regional programme of measures for the marine environment based on the ecosystem approach. Mar Pollut Bull 60(5):642–649

Bäcklin B-M, Moraeus C, Kunnasranta M, Isomursu M (2009) Health assessment in the Baltic grey seal (*Halichoerus grypus*). HELCOM indicator fact sheets 2009. http://www.helcom.fi/BSAP_assessment/ifs/en_GB/cover/

Belkin I (2009) Rapid warming of large marine ecosystems. Prog Oceanogr 81(1–4):207–213

Bignert A, Danielsson S, Nyberg E (2010a) PCB concentrations in fish muscle. HELCOM indicator fact sheets 2010. http://www.helcom.fi/BSAP_assessment/ifs/en_GB/cover/

Bignert A, Danielsson S, Nyberg E (2010b) TCDD-equivalents in herring muscle and guillemot eggs. HELCOM indicator fact sheets 2010. http://www.helcom.fi/BSAP_assessment/ifs/en_GB/cover/

Dippner JW, Vuorinen I, Daunys D, Flinkman J, Halkka A, Köster FW, Lehikoinen E, MacKenzie BR, Möllmann C, Møhlenberg F, Olenin S, Schiedek D, Skov H, Wasmund N (2008) Climate related marine ecosystem change. In: The BACC Author team (eds) Assessment of climate change for the Baltic Sea basin regional climate studies. Springer, Berlin

EC (2000) Directive 2000/60/EC of the European parliament and of the council of 23 October 2000 establishing a framework for community action in the field of water policy (Water framework directive). Official Journal of the European Communities L 327 2000

EC (2001) Directive 2001/81/EC of the European parliament and of the council of 23 October 2001 on national emission ceilings for certain atmospheric pollutants (NEC Directive). Official Journal of the European L 309/22 2001

EC (2008) Directive 2008/56/EC of the European parliament and of the council of 17 June 2008 establishing a framework for community action in the field of marine environmental policy (Marine strategy framework directive). Official Journal of the European Union L 164/19 2008

EC (2009a) Communication from the commission to the European parliament, the council, the European economic and social committee and the committee of the regions concerning the European union strategy for the Baltic Sea region, Brussels, 10.6.2009 COM

EC (2009b) Decision No. 406/2009/EC of the European parliament and of the council of 23 April 2009 on the effort of member states to reduce their greenhouse gas emissions to meet the Community's greenhouse gas emission reduction commitments up to 2020. Official Journal of the European Union L 140/136

Fleming-Lehtinen V, Kaartokallio H (2009) Water transparency in the Baltic Sea between 1903 and 2009. HELCOM indicator fact sheets 2009. http://www.helcom.fi/BSAP_assessment/ifs/en_GB/cover/

Fleming-Lehtinen V, Laamanen M (in press) Long-term changes in Secchi depth and the role of phytoplankton in explaining light attenuation in the Baltic Sea. Estuarine, Coastal and Shelf Science.

Graham LP, Chen D, Christensen OB, Kjellström E, Krysanova V, Meier HEM, Radziejewski M, Räisänen J, Rockel B, Ruosteenoja K (2008) Projections of future anthropogenic climate change. In: The BACC Author team (eds) Assessment of climate change for the Baltic Sea Basin. Regional climate studies. Springer, Berlin

Granskog M, Kaartokallio H, Kuosa H, Thomas DN, Vainio J (2006) Sea ice in the Baltic Sea – a review. Estuar Coast Shelf Sci 70:145–160

Gusev A (2010a) Atmospheric depositions of heavy metals on the Baltic Sea. HELCOM indicator fact sheets 2010. http://www.helcom.fi/BSAP_assessment/ifs/en_GB/cover/

Gusev A (2010b) Atmospheric depositions of PCDD/Fs on the Baltic Sea. HELCOM indicator fact sheets 2010. http://www.helcom.fi/BSAP_assessment/ifs/en_GB/cover/

Helander B, Bignert A, Herrmann C (2009) Predatory bird health – white-tailed sea eagle. HELCOM indicator fact sheets 2009. http://www.helcom.fi/BSAP_assessment/ifs/en_GB/cover/

HELCOM (1974) Convention for the protection of the marine environment of the Baltic Sea area, Helsinki, 22 March 1974, in force 3 May 1980
HELCOM (1988) Declaration on the protection of the environment of the Baltic Sea. Ninth meeting of the Baltic marine environment protection commission, Helsinki, 15 Feb 1988
HELCOM (1992) Convention for the protection of the marine environment of the Baltic Sea area. Adopted in Helsinki, Finland on 9 April 1992, in force 17 Jan 2000
HELCOM (1998) Communiqué of the ministerial session on 26 March 1998. Meeting of the Baltic marine environment protection commission, Helsinki, 26 March 1998
HELCOM (2002) Environment of the Baltic Sea area 1994–1998. Baltic Sea environmental proceedings no. 82B
HELCOM (2006) Development of tools for assessment of eutrophication in the Baltic Sea. Baltic Sea environment proceedings no. 104
HELCOM (2007a) Baltic Sea Action Plan. HELCOM ministerial meeting. Krakow, Poland, 15 Nov 2007
HELCOM (2007b) Climate change in the Baltic Sea area – HELCOM thematic assessment in 2007. Baltic Sea environmental proceedings no. 111
HELCOM (2009a) Biodiversity in the Baltic Sea – an integrated thematic assessment on biodiversity and nature conservation in the Baltic Sea. Baltic Sea environment proceedings no. 116B
HELCOM (2009b) Eutrophication in the Baltic Sea – an integrated thematic assessment of the effects of nutrient enrichment in the Baltic Sea region. Baltic Sea environment proceedings no. 115B
HELCOM (2010a) Hazardous substances in the Baltic Sea – an integrated thematic assessment of hazardous substances in the Baltic Sea. Baltic Sea environment proceedings no. 120B
HELCOM (2010b) Ecosystem health of the Baltic Sea – HELCOM initial holistic assessment. Baltic Sea environment proceedings no. 122
HELCOM (2010c) HELCOM ministerial declaration on the implementation of the HELCOM Baltic Sea Action Plan. HELCOM ministerial meeting, Moscow, 20 May 2010
JMM (2003) Statement on the ecosystem approach to the management of human activities: towards and ecosystem approach to the management of human activities. Annex 5. First joint meeting of the Helsinki and OSPAR commissions (JMM), Bremen, 25–26 June 2003
Knuuttila S (ed) (2009) Waterborne loads of nitrogen and phosphorus to the Baltic Sea in 2006. HELCOM indicator fact sheets 2009. http://www.helcom.fi/BSAP_assessment/ifs/en_GB/cover/
Laine AO (2008) Climate change in the Baltic Sea area: observed and projected changes and possible biological indicators in the marine ecosystem. A draft report of the project "People, Nature and Harbours", Part 1. Metsähallitus, The Natural Heritage Services, Finland
Leppäranta M, Myrberg K (2009) Physical oceanography of the Baltic Sea. Springer, Berlin
Mee LD (2005) Assessment and monitoring requirements for the adaptive management of European's regional seas. In: Vermaat J, Bouwer L, Salomons W, Turner K (eds) Managing European coasts: past present and future. Springer, Berlin, pp 227–237
Nielsen B, Kvaavik C (2007) Pelagic habitat mapping: a tool for area-based fisheries management in the Baltic Sea. BALANCE interim report no. 20
Pyhälä M, Brusendorff AC, Paulomäki H, Ehlers P, Kohonen T (2007) The precautionary principle and HELCOM. In: de Sadeleer N (ed) Implementing the precautionary principle: approaches from the Nordic Countries, EU and USA. Earthscan, London
Sabine CL, Feely RA, Gruber N, Key RM, Lee K, Bullister JL, Wanninkhof R, Wong CS, Wallace DWR, Tilbrook B, Millero FJ, Peng T-H, Kozyr A, Ono T, Rios AF (2004) The oceanic sink for anthropogenic CO_2. Science 305:367–371
Sherman K, Belkin I, Friedland KD, O'Reilly J, Hyde K (2009) Accelerated warming and emergent trends in fisheries biomass yields of the world's large marine ecosystems. Ambio 38(4):215–224
UN (1972) Declaration of the United Nations conference on the human environment, Stockholm, 5–16 June 1972

UN (1992) Declaration on environment and development. The United Nations conference on environment and development, Rio de Janeiro, 3–14 June 1992
UN (2002) Plan of implementation of the 2002 world summit on sustainable development, Johannesburg, 26 Aug–4 Sep 2002
UNECE (1999) Protocol to the 1979 convention on long-range transboundary air pollution to abate acidification, eutrophication and ground-level ozone, Gothenburg, 30 Nov 1999
Wulff F, Savchuk OP, Sokolov A, Humborg C, Mörth C-M (2007) Management options and effects on a marine ecosystem: assessing the future of the Baltic. Ambio 36(2–3):243–249

Chapter 3
Some Aspects of the Baltic Sea Marine System

Anders Omstedt

3.1 Introduction

The development within science today is to integrate knowledge from many different disciplines addressing both environmental and climate change issues (e.g. BALTEX Phase II). As a part of this strategy, Earth System Models are now being developed. This implies that much of the coming model results need to be understood from a systems point of view. However, system understanding all relies on good process understanding. We may expect that process understanding will develop through improved knowledge within our special disciplines, such as air-sea interactions within meteorology, mixing processes within oceanography, acidification within chemistry and ecosystems within biology. However, understanding of how the processes interact and influence the state of the sea is strongly needed. This will require a deeper understanding of the whole system, challenging the interdisciplinary talents of scientists. The objective of the present chapter is to give an overview of the Baltic Sea as a marine system by illustrating some important processes that need to be involved in such an understanding.

The chapter starts by demonstrating conservation principles and governing equations, forming the base for our system understanding. Some physical aspects of marine waters are then discussed and the importance of simplifications is stressed. The traditional approach in oceanography is then to learn about ocean circulation through water mass analysis. Water masses can often be understood as pools of water which are dynamically controlled by different physical transport mechanisms such as dense bottom currents, intrusion layers, strait flows and selected withdrawal (water layers involved in the outflow). We then discuss turbulence which strongly influences the Baltic Sea both at the surface and in deep waters. The water, salt and heat balances of the Baltic Sea, which was a major focus of BALTEX Phase I, are discussed. Then, the Baltic Sea nutrient balances and some primary production aspects are considered.

A. Omstedt (✉)
Department of Earth Sciences, University of Gothenburg, 460,SE-405 30 Göteborg, Sweden
e-mail: Anders.Omstedt@gvc.gu.se

M. Reckermann et al. (eds.), *Climate Impacts on the Baltic Sea: From Science to Policy*, Springer Earth System Sciences, DOI 10.1007/978-3-642-25728-5_3,

The close link between nutrients, carbon and biological processes (through carbon dioxide) opens a discussion about the acid–base (pH) balance of marine waters. Finally, we close with a short discussion about climate change and make reference to some of the long-time data sets available in the Baltic Sea region.

3.2 Two Illustrative Model Examples

Field measurements of temperature and other water properties are often scarce. In aquatic studies with few direct observations, a combination of observations and models is therefore often needed. Figure 3.1 shows how the results of a coastal sea model can reproduce sea surface temperature (SST) in various parts of the Baltic Sea. Models can be validated over periods and regions for which we have a high measurement frequency. When measurements become sparse, for example in Bothnian Bay (Fig. 3.1), we have only a rough idea of the environmental conditions of that time. From the figure it is clear that accurate trends cannot be detected based on observations alone; instead a combination of observations and modelling are needed to allow changes to be detected.

The potential to extrapolate information (in time and space) is a good reason why we use models: they help us understand processes outside our observational knowledge. An example related to the Baltic Sea water balance is given below. Over the last 100 years, the freshwater inflow has hovered around a mean of 15,000 m^3/s. What would happen if the freshwater supply increased? The curves in Fig. 3.2 indicate that if the freshwater inflow was to be tripled, the Baltic Sea would be transformed into a freshwater sea. Can we rely on this result, and how are we supposed to know? The calculations indicate that the Baltic Sea is sensitive to variations in freshwater inflow, yet it is highly unrealistic to conceive such a drastic increase of several hundred percent.

The calculations indicate that the Baltic Sea will remain a brackish sea in the future. Still, these results could be challenged as new knowledge becomes available. This example has interesting implications for the model results which lie outside of the observed range. When using models, we of course need to be careful with extrapolations, as they are unsupported by direct validation data.

3.3 Conservation Principles and Governing Equations

Conservation equations can be formulated for most aquatic properties (ϕ) and for transient three-dimensional problems. The general differential equation is:

$$\frac{\partial}{\partial t}\phi + U_i\frac{\partial \phi}{\partial \chi_i} = S_\phi \text{ where } i = x, y, z$$

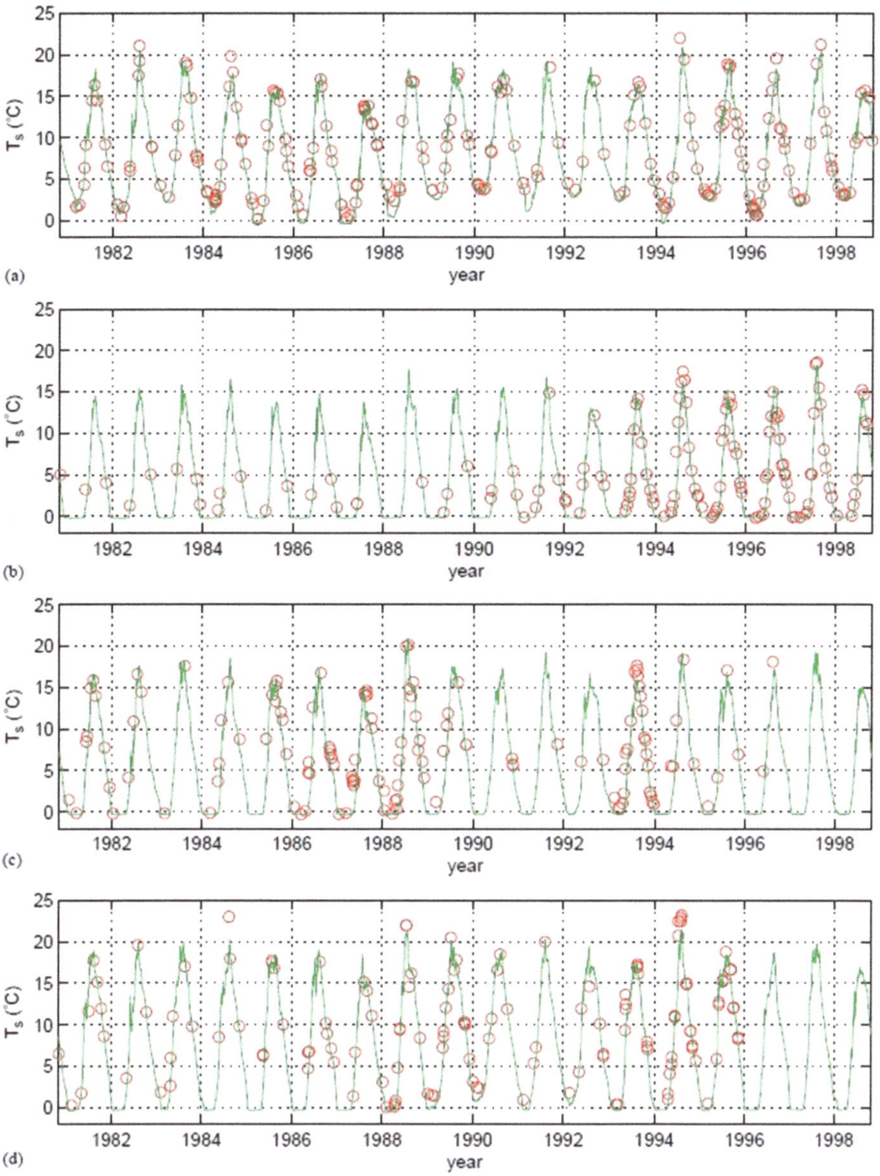

Fig. 3.1 Modelled (*fully drawn lines*) and observed (*circles*) sea surface temperatures in (**a**) Eastern Gotland Basin, (**b**) Bothnian Bay, (**c**) Gulf of Finland, and (**d**) Gulf of Riga (Omstedt and Axell 2003)

Here, ϕ could stand for momentum, temperature, salinity, oxygen and nutrients, U_i stands for velocity, with the indices indicating velocity components in horizontal and vertical directions, and S_ϕ is the source/sink term related to the properties considered. The coordinates in space are denoted x, y and z, while t is the coordinate

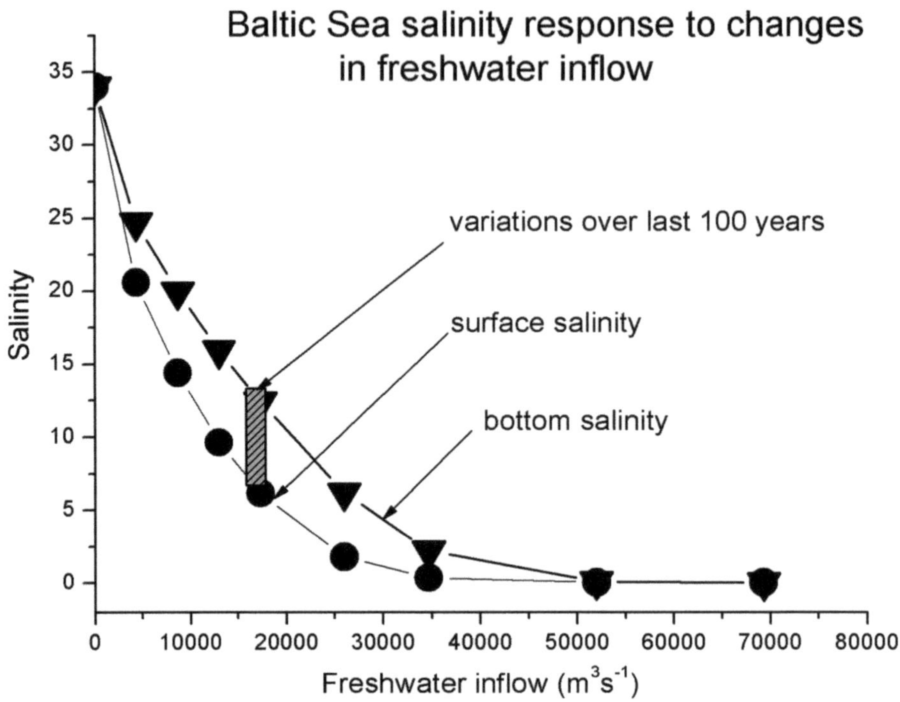

Fig. 3.2 Model calculations indicating how salinity in the central Baltic Sea can vary with variations in freshwater inflow from rivers and precipitation. The hatched field represents the typical observed variation over the last 100 years (Redrawn from Omstedt and Hansson 2006a, b)

in time. As geophysical flows are typically turbulent, but also include waves, their properties can be divided into mean, wave, and fluctuation or turbulent parts, and simplifications can then be made based on scale analysis. If we neglect wave motion, the division can be formulated as

$$U_i = \bar{U}_i + U'_i$$
$$\phi_i = \bar{\phi}_i + \phi'_i$$

where $\overline{U}_i$, $\overline{\phi}_i$ represent the mean velocity and mean property, and U'_i, ϕ'_i represent the corresponding fluctuating parts. The equations derived from the conservation principles using the Reynolds method of averaging and the eddy viscosity concept read

$$\frac{\partial}{\partial t}\overline{\phi} + \overline{U_i}\frac{\partial\overline{\phi}}{\partial x_i} = \frac{\partial}{\partial x_i}\left(\Gamma_\phi \frac{\partial}{\partial x_i}\overline{\phi}\right) + \overline{S_\phi}$$

where Γ_ϕ includes the sum of molecular and turbulent diffusion processes and has the dimension [UL], where U and L are the velocity and length scales respectively. The terms of the conservation equation, from left to right, represent property changes: in time, due to advection, due to turbulent diffusion, and due to sources/ sinks. The geophysical flow equations are described in detail in Cushman-Roisin and Becker 2011.

Source/sink terms in the momentum equation are pressure and gravity. The Coriolis effect due to Earth rotation is a part of acceleration but can also be treated as a source term. The source term in the temperature equation is the sun radiation penetrating the water body.

An important aspect of the general conservation equation is that it can form the basis for developing general equation solvers. Based on this equation, one can derive the corresponding discretized equation through integration over a control volume. The general discretization equation can then be formulated as

$$a_P\theta_P = a_W\theta_W + a_E\theta_E + a_S\theta_S + a_N\theta_N + a_B\theta_B + a_T\theta_T + S_\theta$$

where we now use θ to represent the finite volume form of ϕ. The indices represent the calculations in position P determined from information from the west (W), east (E), south (S), north (N), bottom (B), and top (T) nodal points. Thus, the solution in a finite volume is closely linked to the surrounding grid points and will also change in time. Special care needs to be taken at the boundary layers and in the start of the calculation. This type of equation forms the basis of several general equation solvers (Versteeg and Malalasekera 1995) which effectively solve fluid flow problems.

3.4 Sea Water Properties and Ice

A description of a fluid needs to relate its density to its state variables, a relationship called the equation of state. This equation is generally a function of temperature, salinity, and pressure (Gill 1982, Appendix 3), but for coastal seas, one can often neglect pressure and use the following approximation

$$\rho = \rho_0\left(1 - \alpha\left(T - T_{\rho m}\right)^2 + \beta(S - S_0)\right)$$
$$T_{\rho m} = 3.98 - 0.22S$$

where T and S are temperature and salinity, ρ_0 is reference density, $T_{\rho m}$ the temperature of maximum density, S_0 reference salinity, and α and β are thermal and salinity expansion/contraction, respectively. Typical values for brackish water are $\rho_0 = 1000$ ($m^3\ s^{-1}$), $S_0 = 0$, $\alpha = 5.10^{-6}$ ($°C^{-2}$), and $\beta = 8.10^{-4}$. Pressure is not included in the equation of state, indicating that water is incompressible. This is often a good simplification but does not hold for processes such as sound propagation.

The freezing temperature of sea water is

$$T_f = -0.0575S + 0.0017S^{1.5} - 0.0002S^2 - 0.00753P$$

where P is the pressure in bars (one can neglect the pressure term for shallow seas). Brackish water is often defined as saline water that has a freezing point below the temperature of maximum density. The upper salinity limit is 24.7 and the lower limit is estimated to be 0.5 (Leppäranta and Myrberg 2009).

In the case of freezing water, ice properties need to be considered. Ice forms a thin, rigid but fragile layer over the water body that dramatically changes the heat, momentum, and gas exchanges between the atmosphere and water. Ice forms in two different ways: Under calm conditions when the water surface is slightly super cooled and after atmosphere seeding, ice crystals start growing horizontally into large ice crystals. When the horizontal space is filled up, the crystals start growing vertically (in seawater, only pure ice crystals grow and the salinity leaks out along the crystal boundaries). These ice crystals form columnar ice that, together with snow, is often seen in sheltered lakes and inland waters. Under windy conditions accompanied by super cooling, small ice crystals become mixed into the water column forming frazil ice. Under open water conditions, all the heat that escapes the cold water surface is used for ice production, resulting in huge amounts of frazil ice. The frazil ice is later transformed to grease ice and pancake ice, which often constitutes a base for sea ice formation.

When a thin ice layer has formed, winds and currents cause it to drift. During free ice drift, the ice moves at approximately 2–3% of the wind velocity and 20–30° to the right of the wind direction, due to the Coriolis effect (in the Northern Hemisphere). During onshore winds or when the ice concentration is high, the plastic behaviour of ice starts to influence the drift. This reduces the velocity, but at a certain ice pressure, the ice breaks and starts forming ridged ice.

3.5 Simplifications Needed in Model Design

Models are important tools for summarizing and testing our understanding. They are often based on a number of different conservation equations which are non-linear, so it is often impossible to find analytical solutions except in simple flow cases. Instead, we must rely on numerical methods to solve the equations. Before making any model design, we must systematically reduce the problem complexity by means of proper simplification. The first step is then to carefully formulate the problem. We may then look into a sea chart and identify the water volume and divide it, guided by observation, into dynamically relevant processes. Hypsographic (area-depth distribution) curves give important information of the water volumes available at different depths, while bathymetric charts tell us where narrow straits, channels, and canyons are and where natural sub-basins occur (Fig. 3.3). Knowledge of forcing functions (e.g., meteorological and hydrological conditions) also provides a good basis for model design. As water circulation is often of key importance, hydrodynamic equations should be simplified by identifying the scales of motion.

The use of important dimensionless numbers can help us in searching for simplifications. Some of the main dimensionless numbers used in geophysical

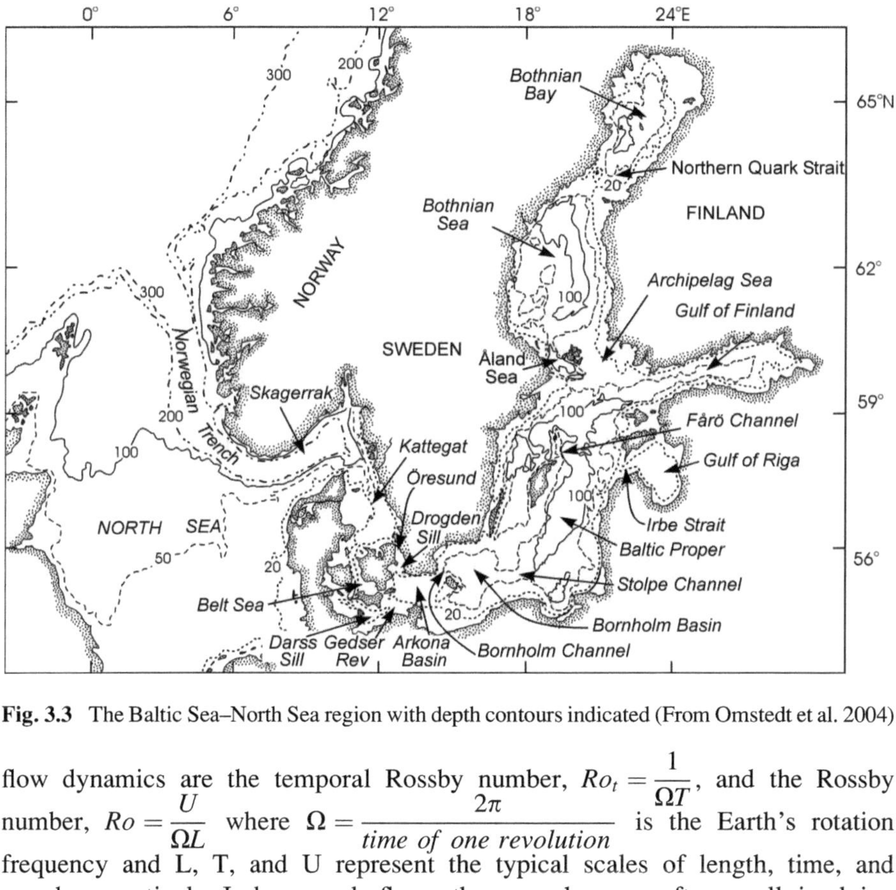

Fig. 3.3 The Baltic Sea–North Sea region with depth contours indicated (From Omstedt et al. 2004)

flow dynamics are the temporal Rossby number, $Ro_t = \frac{1}{\Omega T}$, and the Rossby number, $Ro = \frac{U}{\Omega L}$ where $\Omega = \frac{2\pi}{time\ of\ one\ revolution}$ is the Earth's rotation frequency and L, T, and U represent the typical scales of length, time, and speed, respectively. In large-scale flows, these numbers are often small, implying that the acceleration terms are small.

The importance of friction can be estimated from the horizontal, $Ek_h = \frac{\mu_t}{\rho \Omega L^2}$, and vertical, $Ek_v = \frac{\mu_t}{\rho \Omega H^2}$, Ekman numbers. Outside the boundary layers, the Ekman numbers are small and friction effects can be neglected. The implications of these numbers can be easily understood if we examine the equations governing geophysical flows, considering only the x-dimension and with estimated scales under the different terms:

$$\begin{array}{cccccccccc}
\frac{\partial}{\partial t}\overline{u} & +\overline{u}\frac{\partial \overline{u}}{\partial x} & +\overline{v}\frac{\partial \overline{u}}{\partial y} & +\overline{w}\frac{\partial \overline{u}}{\partial z} & -f\overline{v} & = -\frac{1}{\rho_0}\frac{\partial p}{\partial x} & +\frac{\partial}{\partial x}\left(\Gamma_h \frac{\partial \overline{u}}{\partial x}\right) & +\frac{\partial}{\partial y}\left(\Gamma_h \frac{\partial \overline{u}}{\partial y}\right) & +\frac{\partial}{\partial z}\left(\Gamma_z \frac{\partial \overline{u}}{\partial z}\right) \\
\frac{U}{T} & \frac{U^2}{L} & \frac{U^2}{L} & \frac{WU}{L} & \Omega U & \frac{P}{\rho_0 L} & \frac{\mu_h U}{\rho_0 L^2} & \frac{\mu_h U}{\rho_0 L^2} & \frac{\mu_v U}{\rho_0 H^2}
\end{array}$$

Here, p is the pressure and f the Coriolis parameter. By dividing the estimated sizes beneath the equation by ΩU, we can easily identify the different nondimensional numbers given above. For example, if $Ro_t << 1$, the transient term can be neglected relative to the rotation part.

The incompressible approximation for water leads to the important assumption of divergence-free motion (represented by the continuity equation). That is, if the motion is slow, any imbalance in horizontal motion will be reflected in vertical motion:

$$\frac{\partial \overline{u}}{\partial x} + \frac{\partial \overline{v}}{\partial y} + \frac{\partial \overline{w}}{\partial z} = 0$$

An important analytical simplification is that large-scale circulation is often in geostrophic and hydrostatic balance, implying a balance between the Earth's rotation and the horizontal pressure gradients in the horizontal dimensions, and between gravity and the vertical pressure gradients in the vertical dimension. In this case, the dimensionless numbers Ro_t, Ro, Ek_h, Ek_v all are much less than 1.

At the surface and bottom boundary layers (i.e., $Ek_v \approx 1$), one can also find analytical solutions to the boundary equations that demonstrate the presence of typical Ekman boundary layer properties. For the surface boundary layer, the net mass transport is perpendicular to the wind direction and to the right on the northern hemisphere. The analytical solution for the mass transport in the bottom boundary layer demonstrates a transport to the left of the geostrophic flow direction.

If Ro_t; R_o; Ek_h; $Ek_v << 1$, several important analytical aspects of the fluid flow can be derived. For example, the Taylor–Proudman theorem states that the horizontal velocity field has no vertical shear and that the flow cannot flow across changes in bottom topography; instead, all motions follow the depth contours. The bathymetric charts then give us important information about the circulation, with implications for the division of the water body into dynamic regions.

For time-dependent geostrophic flow, one can derive a useful expression for the vorticity dynamics; with $Ro_t Rk_v << 1$ and $Ro_t \leqslant 1$, the change in potential vorticity needs to be conserved, as follows:

$$\frac{d}{dt}\left(\frac{f + \frac{\partial \overline{v}}{\partial x} - \frac{\partial \overline{u}}{\partial y}}{H}\right) = 0$$

For a change in for example water depth (H), the equation implies that the fluid flow will change its vorticity.

In geometric constrictions such as narrow straits, the non-linear terms become important and these terms can be balanced by pressure effects. For such a flow (i.e., constant density, steady state, no friction, and no rotation), a useful simplification is provided by the Bernoulli's equation:

$$p + \frac{1}{2}\rho U^2 + \rho g H = \text{constant.}$$

In stratified fluids, the Froude number which is the ratio of flow speed to the phase speed of long internal gravity waves, i.e. $Fr = \frac{U}{NH}$, where N is the stratification frequency or Brunt–Väisälä frequency ($N^2 = -\frac{g}{\rho_0}\frac{d\rho}{dz}$), gives us some useful information. The rule is that if $Fr \leqslant 1$, then the stratification effects are important. The length scale at which stratification and rotation become equally important is called the Rossby radius of deformation (i.e. $L_{Ro} = \frac{NH}{\Omega}$).

When the fluid is stratified, one can simplify the equations by assuming wave solutions. If we assume a wave solution with no space limitation, $w = w_0 e^{i(lx+my+nz-\omega t)}$ (note that $e^{i\phi} = \cos\phi + i\sin\phi$), one can demonstrate that the internal wave frequency is limited by the Coriolis frequency and the stratification frequency (i.e. $f \prec \omega \prec N$). Such waves, called inertia-gravity waves, are oscillating in nature and typical of open sea conditions. The wave number and frequency are denoted by

$$n = \frac{2\pi}{L_x}, m = \frac{2\pi}{L_y}, n = \frac{2\pi}{L_z} \text{ and } \omega = \frac{2\pi}{T}.$$

where L_x, L_y, and L_z are the wavelength in x, y and z directions, and T is the wave period.

If we instead assume a wave solution, i.e. $w = w_0 e^{i(my+nz-\omega t)-lx}$, with one coast along $x = 0$, the results indicate that the waves have no lower frequency limit ($\omega \prec N$). These waves, called coastal-trapped waves, move along the coast with the coast to the right of the wave propagation and often at a low frequency. In semi-enclosed basins, one may thus expect to have coastal-trapped waves along the coastlines without lower limits on the wave frequency; in the open part of such basins, one may encounter oscillating currents bound by the stratification and the Coriolis frequencies.

The analytical solutions have important implications for any model design. However, sooner or later we must face the fact that we need to solve the simplified equations using numerical methods. Numerical models need to apply grids that have finite scales and the equations can be solved only within a limited resolution. Dynamic features larger than the grid domain must then be prescribed, while features below the grid size must be parameterized. If we were to construct a numerical model of, for example, coastal upwelling (with typical spatial and time scales of $L = 10$ km, $H = 100$ m, and $T = 10$ days), we would need a domain at least 10 times as large as the upwelling in both space and time. To resolve the motions, a grid size of at least 1/10 of the typical spatial and time scales to be resolved is needed. Given these considerations, we need to prescribe motion on a horizontal scale of 100 km and a vertical scale of 1,000 m, as well as computer time of 100 days. Processes such as geostrophic eddies, thermohaline circulation, tides, and climate effects (Fig. 3.4) must therefore be prescribed. The resolved motion includes the front dynamics, but we should be aware that errors in numerical models could be of several types, such as discretization, iteration, and rounding errors. Numerical tests, conservation checks, and validation using ocean observations are therefore needed to be on the safe side. Parameterized motions are those at grid scale, in this case at a horizontal scale of 1 km, a vertical scale of 10 m, and a time interval of 1 day. All motions at these spatial and time scales need to be parameterized, including turbulence, surface waves, Langmuir circulation, swell, turbulent effects from shipping, and breaking waves.

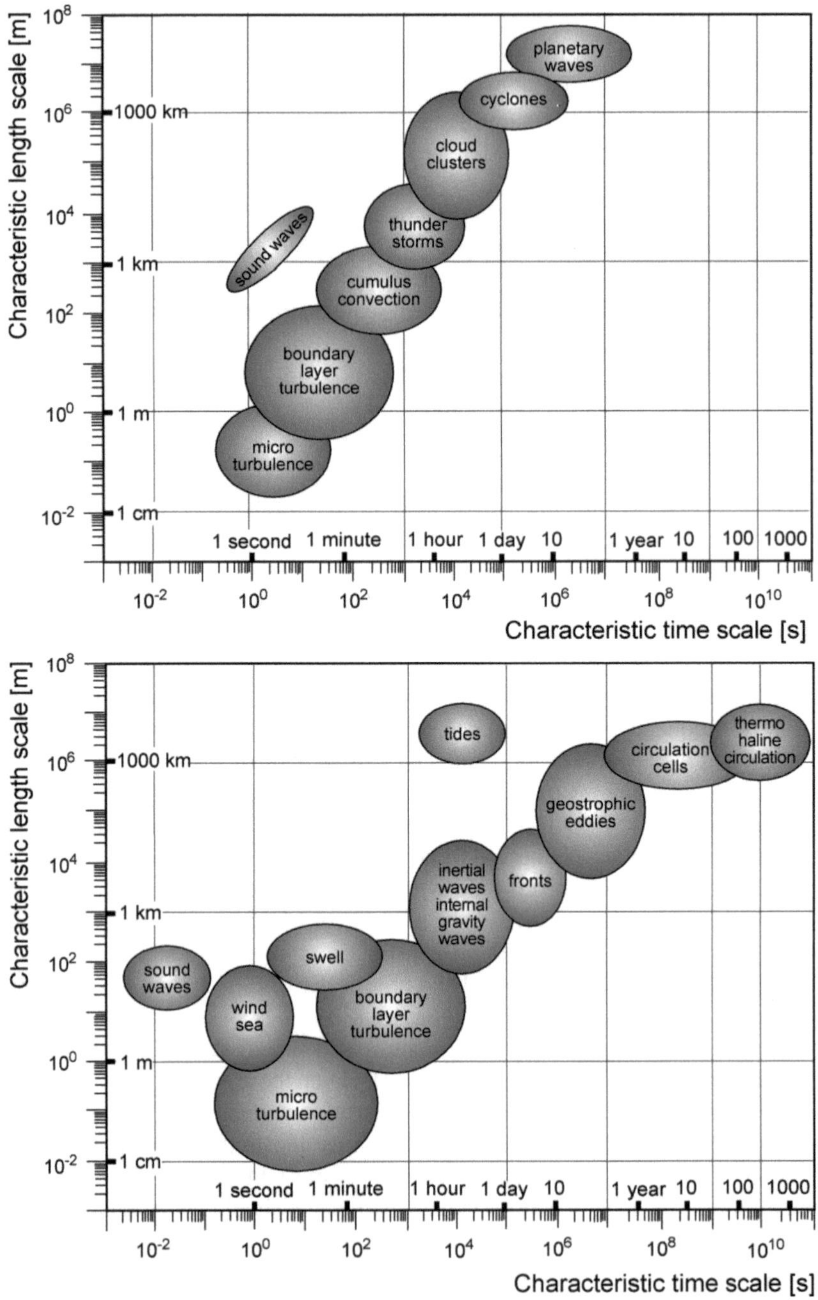

Fig. 3.4 Spatial and temporal scales of some atmospheric and ocean processes (From von Storch et al. (2004))

3.6 Water Masses and Water Pools

Water mass and trace analysis have been the two basic methods for determining large-scale ocean circulation (e.g. Aken 2007). Based on temperature and salinity measurements, and plotting these variables in a T–S diagram, we may obtain important information about ocean circulation and mixing processes. For example, when two water masses with different T and S are mixed, the new water mass is connected by a straight line between the water masses. Figure 3.5 depicts the long-term T–S structure of some sub-basins of the Baltic Sea. Starting from the deep north-western Gotland Basin, the T–S structure indicates three different water masses: (1) the surface layer with large temperature variations and small salinity

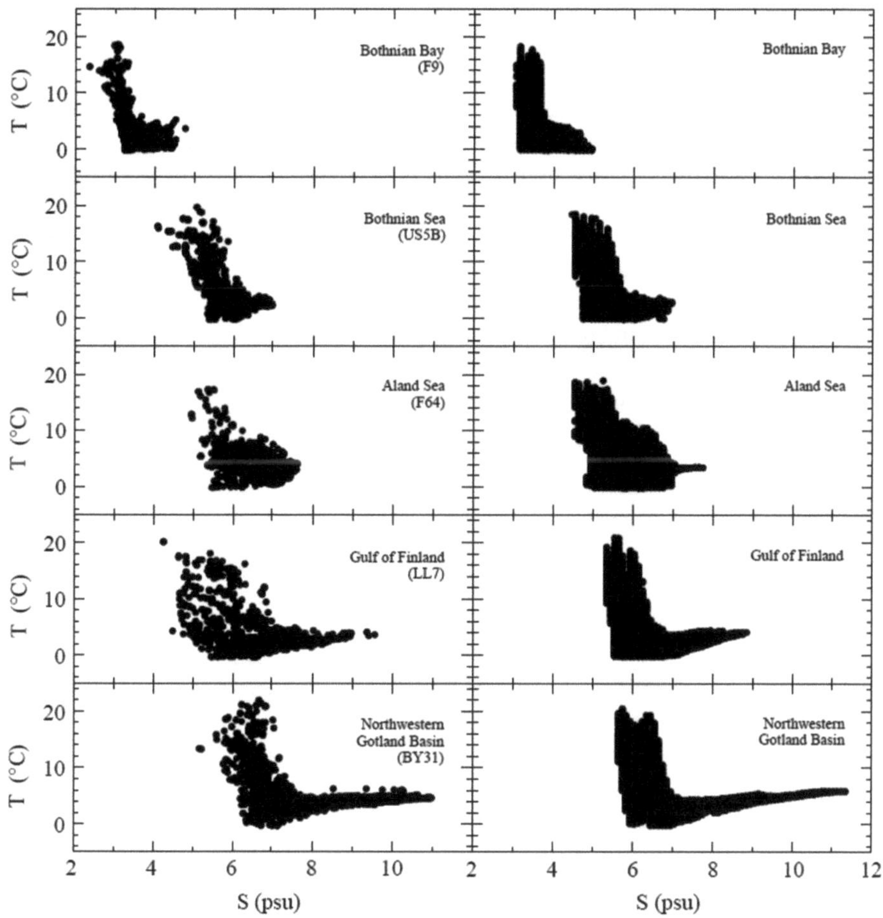

Fig. 3.5 Observed (*left*) and calculated (*right*) long-term T–S structure in the Northern Baltic Sea (From Omstedt and Axell 2003)

variation, (2) the halocline layer with quite small temperature and salinity variations, and (3) the deep-water layer with small temperature variations but larger salinity variations. In the Åland Sea, Bothnian Sea, and Bothnian Bay water masses, the surface water is diluted by river runoff, and the deep-water masses are cut off from such dilution by the presence of sill regions.

The water masses often form pools of water under specific dynamic control. Analysing the dynamics underlying the filling and emptying of such pools can often lead to new insights and interesting model simplifications (Stigebrandt 2001); for example, we can identify a Kattegat surface-water pool and a Bornholm Basin deep-water pool.

3.7 Strait Flows

Inshore–offshore water exchange and exchange between sub-basins are often controlled by geometric constrictions such as sills and straits (Fig. 3.6). Straits are narrow, shallow regions, and a basic understanding of strait flow dynamics (including mixing) and the proper parameterization of strait flows is needed. In this context, one can generally speak of barotropic and baroclinic flows, associated with sea level and density changes, respectively.

The dynamics through the Sound and the Belt Sea are very transient, oscillating from 0 to 100,000 m^3/s in both directions, so direct measurements are quite difficult to make. As the strait flows in the Sound and the Belt Sea are mainly driven by differences in sea level across the entrance area, the in- and outflows have so far mainly been calculated from observations of sea level or from models (Fig. 3.7).

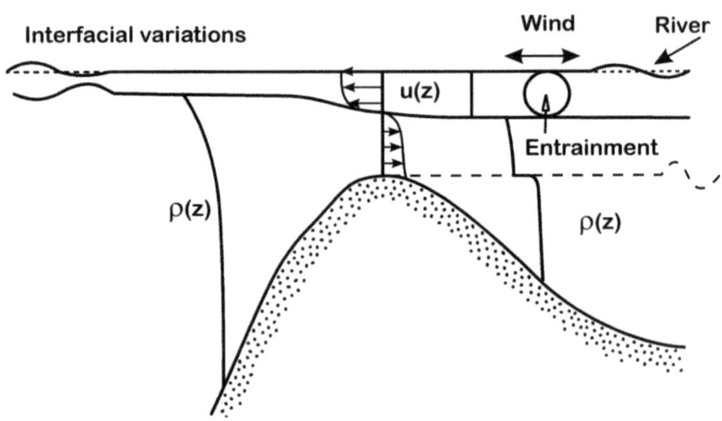

Fig. 3.6 Exchange in an inshore-off shore region with a sill (Redrawn from Green 2004)

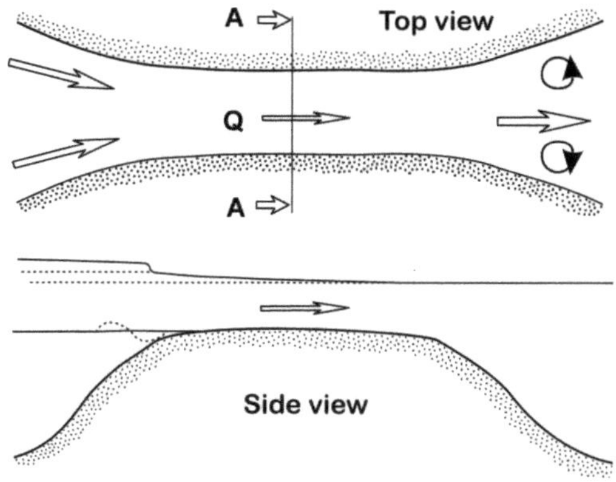

Fig. 3.7 Barotropic exchange through a shallow channel (Redrawn from Green 2004)

This complex exchange can be modelled according to volume conservation and a simple barotropic strait flow model (Stigebrandt 2001):

$$A_s \frac{dz_s}{dt} = Q_b + (P - E)A_s + Q_r$$

$$Q^2{}_b = \frac{1}{c_s} \Delta z$$

Here, A_s is the surface area of the Baltic Sea, z_s the water level of the Baltic Sea, Δz the sea level difference between the Baltic Sea and the Kattegat, Q_b the barotropic in- and outflows through the Baltic entrance area, c_s a strait-specific constant, P and E the precipitation and evaporation rates, and Q_r the river runoff.

Fresh water input from land often results in surface layers that mix with the surrounding coastal waters and denser, deeper waters that flow inshore. In straits wider than the Rossby radius of deformation, a geostrophic baroclinic flow can be assumed:

$$Q_g = \frac{g(\frac{\rho_2 - \rho_1}{\rho_0})H_1^2}{2f},$$

where H_1 represents the depth of a mixed surface layer of density ρ_1 overlying a deeper layer of density ρ_2. In narrow straits, the concept of baroclinic controls can be used (e.g. Stigebrandt 2001). Of special interest is the concept of maximum baroclinic transport capability in the Northern Kvark Strait, connecting the Bothnian Bay to the Bothnian Sea, as suggested by Stigebrandt (2001) and later validated through model simulations by Omstedt and Axell (2003) and studied in field by Green et al. (2006).

In modelling the exchange between inshore and coastal water, a Bernoulli equation strait flow model can often be used.

3.8 Intrusions and Selected Withdrawal

Depending on its density, water may form surface brackish layers like in the Kattegat, or dense bottom layers like in the Baltic Proper (Fig. 3.8).

These surface or bottom water masses are filled by dense- or light-water currents and by mixing processes. Sometimes these dynamics are very transient, with the formation of pulses and eddies indicating transient dynamics (Fig. 3.9) and strong mixing (Piechura and Beszczynska-Möller 2004). Selected withdrawal dynamics may also be considered when examining the outflow levels. One such example is the Sound, where Mattsson (1996) estimated that the water flowing into the Baltic Sea comes from levels below the Drogden Sill depth.

3.9 Turbulence

All flows become dynamically unstable above a certain Reynolds number, i.e., $Re = \frac{UL}{\nu}$, where ν is the kinematic viscosity. At low Reynolds numbers, the flow is laminar but at high values, the flow is turbulent. A turbulent flow is chaotic and random including a whole spectrum of eddies of different sizes. From the energy-

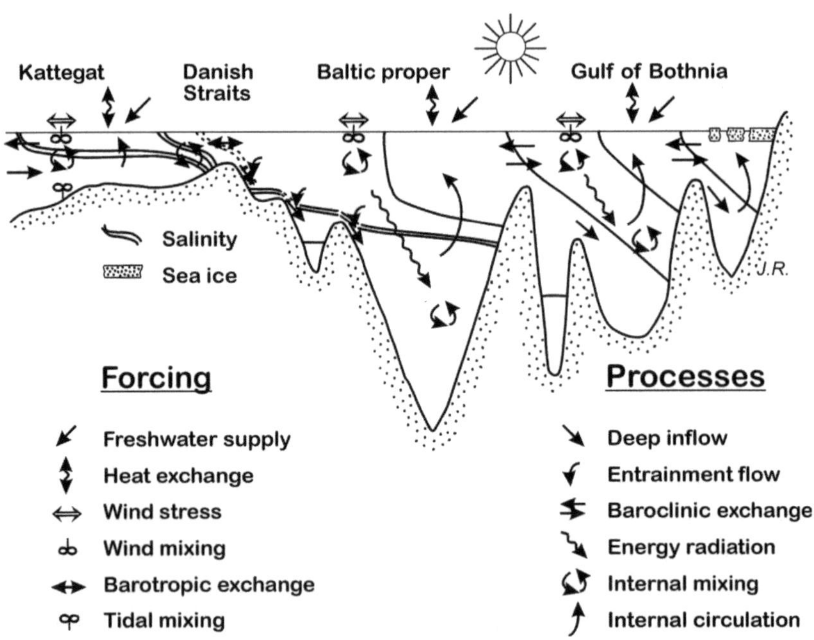

Fig. 3.8 Schematic of processes and forcing mechanisms in the Baltic Sea (Redrawn from Winsor et al. 2001, 2003)

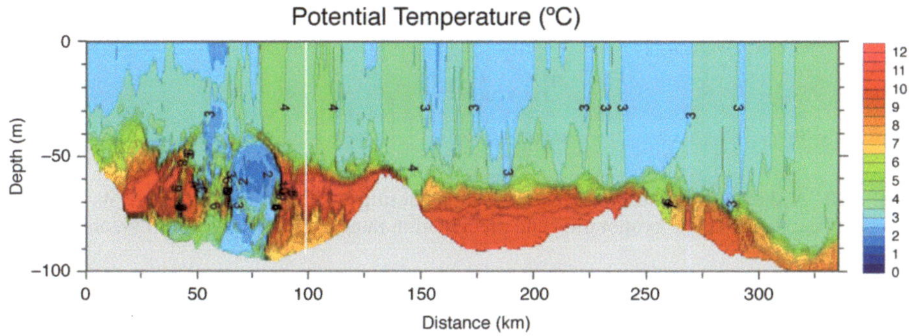

Fig. 3.9 Measured temperatures in the Bornholm Basin (*left*) and the Stolpe Channel (*right*) during January 2003. The *left* illustrates a cold water inflow entering into warm bottom water. In the *right* figure, one can notice a Stolpe Sill overflow with unusual warm temperatures (From Piechura and Beszczynska-Möller 2004)

containing part of this spectrum, the eddies are transformed into smaller eddies, through the inertial sub range and into the dissipative range where the kinetic energy is dissipated as heat. The energy to feed the turbulence comes from the mean currents or from convection, and the size of these energy-containing eddies is often determined by the geometry of the flow. In our mathematical formulation we can represent the turbulent processes using an effective eddy diffusion formulation of momentum, as follows:

$$\Gamma_{\rho u_i} = \frac{\mu}{\rho} + \frac{\mu_t}{\rho}$$

Here, μ and $\frac{\mu}{\rho}$ represent the dynamic and kinematic viscosities, respectively. The dimension of the kinematic viscosity is [UL]. The characterization of eddy viscosity thus requires one equation for the turbulent velocity scale and another for the turbulent length scale. These two equations need to model all processes that both generate and reduce turbulence. Typical processes that generate turbulence are mean current shear, cooling surface water above temperature of maximum density, freezing sea ice, and breaking surface or internal waves. In stratified fluids, turbulent energy is taken to increase the potential energy. Reduction of turbulent energy is also taking part at interfaces (e.g. the air–water, ice–water, and water-bottom interfaces) and during dissipation.

3.10 Water and Salt Balances

The water and heat cycles are the heart of the climate system and thus it is important to understand the interconnections between these cycles. Estimating the different components of the Baltic Sea water balance is therefore crucial (Fig. 3.10), and was the main objective of BALTEX Phase I.

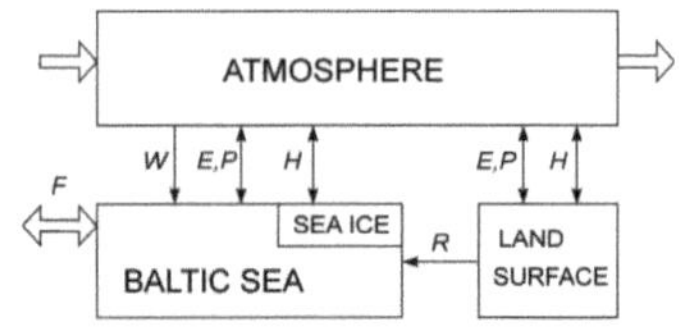

Fig. 3.10 A sketch of the main research questions during BALTEX Phase I. Here W indicates wind, H heat flux, E and P evaporation and precipitation rates, F inflows and outflows, and R river runoff

Starting from the volume conservation principle, we can formulate the water balance equation as (Omstedt et al. 2004)

$$A_s \frac{dz_s}{dt} = Q_i - Q_o + (P - E)A_s + Q_r + Q_g$$

where A_s is the surface area of the Baltic Sea, z_s the water level of the Baltic Sea, Q_i and Q_o the in- and outflows through the Baltic entrance area, P and E the precipitation and evaporation rates, Q_r the river runoff, and Q_g the groundwater inflow. The left term in the equation is the change in water storage, which can be important in short-term estimations of the water balance.

The major water balance components of the Baltic Sea are in- and outflows at the entrance area, river runoff, and net precipitation (Fig. 3.11).

The long-term mean water outflow from the Baltic Sea is estimated to be 80,000 m^3/s, with an inter annual variability of $\pm$10,000 m^3/s. River runoff under current climatic conditions (i.e., the last 100 years) is estimated (the Belt Sea and Kattegat excluded) to be 14,000 m^3/s, with an inter annual variability of $\pm$4,000 m^3/s. Net precipitation over the Baltic Sea under current climatic conditions (i.e., the last 100 years) is estimated to be 1,500 m^3/s, with an interannual variability of $\pm$1,000 m^3/s. The accuracy of the different terms is unknown, but the mean error of the net water balance on a decadal time scale has been estimated to be approximately 600 m^3/s (Omstedt and Nohr 2004).

If the water balance is known, it is possible to calculate the salt balance, which, based on the conservation principals, read as follows:

$$\frac{dVS}{dt} = S_i Q_i - SQ_o - S((P - E)A_s + Q_r),$$

where V and S are the water volume and salinity of the semi-enclosed basin and S_i is the salinity of the inflowing water. Salinity is closely linked to the water balance and forms often the basis of many budget calculations. A useful integral property is the mean salinity (vertically and horizontally integrated) or the fresh water content. The annual mean salinity of the Baltic Sea has during last century varied around 7.6, with a variability of about $\pm$0.5 (Winsor et al. 2001, 2003).

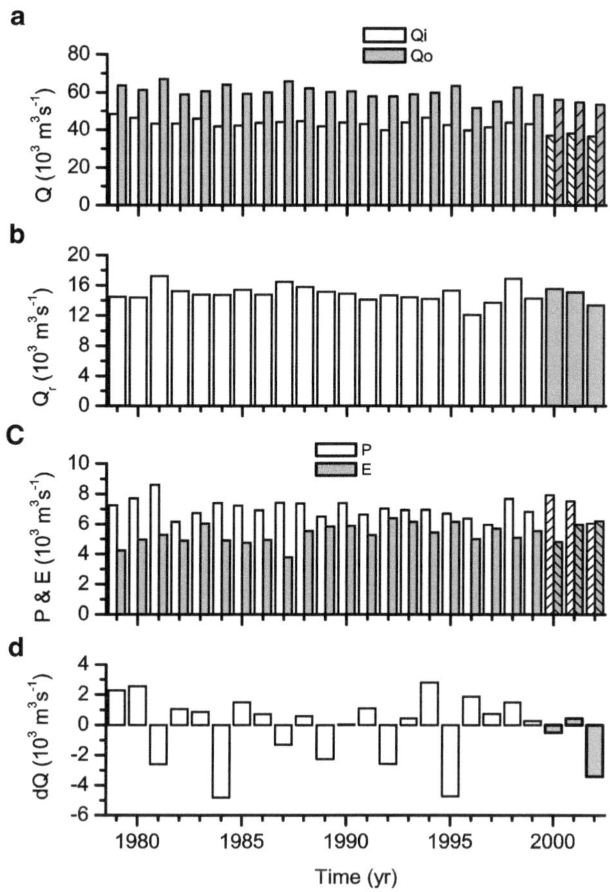

Fig. 3.11 Baltic Sea (excluding the Kattegat and the Belt Sea) annual mean (**a**) inflows and outflows, (**b**) river runoff, (**c**) net precipitation, and (**d**) net volume change (Omstedt and Nohr 2004)

3.11 Heat Balance

From heat conservation principles, we can formulate the heat balance equation for a semi-enclosed sea area, according to Omstedt and Rutgersson (2000), as

$$\frac{dH}{dt} = (F_i - F_o - F_{loss})A_s,$$

where $H = \int\int \rho c_p T dz dA$ is the total heat content of the sea, F_{i} and F_{o} are the heat fluxes associated with in- and outflows, and F_{loss} is the total heat loss to the atmosphere (note that the fluxes are positive when going from the water to the atmosphere). F_{loss} is formulated as

$$F_{loss} = (1 - A_i)(F_n + F_s^o) + A_i(F_w^i + F_s^i) + F_r + F_g,$$

where

$$F_n = F_h + F_e + F_l + F_{prec} + F_{snow}.$$

The various terms of equations are denoted as follows: $A_{\rm i}$ is the ice concentration, F_h the sensible heat flux, $F_{\rm e}$ the latent heat flux, $F_{\rm l}$ the net long-wave radiation, $F_{\rm prec}$ and $F_{\rm snow}$ the heat fluxes associated with precipitation in the form of rain and snow, respectively, $F_{\rm s}^{\rm o}$ the sun radiation to the open water surface, $F_w^{\ i}$ the heat flux from water to ice, $F_s^{\ i}$ the sun radiation through the ice, and $F_{\rm r}$ and $F_{\rm g}$ the heat flows associated with river runoff and groundwater flow, respectively.

Over the long term, the Baltic Sea is almost in thermodynamic balance with the atmosphere. The dominant fluxes, in terms of annual means, are the sensible heat, latent heat, net long-wave radiation, solar radiation to the open water, and heat flux between water and ice (Fig. 3.12). The coupling between the heat and water balances is through the latent heat flux, which together with precipitation determines the net precipitation.

The accuracy of modelling the different fluxes is unknown, but the net heat balance on a decadal time scale can be modelled with a mean error of some Wm^{-2} (Omstedt and Nohr 2004). A useful integral property is the heat content of water (or the vertically and horizontally integrated water temperature) and particularly the change of heat content that is directly connected to the heat fluxes. Surface temperatures are often a poor measure of the heat content and may even not be able to measure the accumulation of heat in the water, as other fluxes may compensate the net effect. Pielke (2003) therefore suggested that the storage of heat content and its changes over time should be the focus in international climate monitoring programmes.

3.12 Nutrient Balance

The biogeochemical modelling of the ocean involves many processes. A description of a marine system is therefore often divided into physical, chemical, and biological components. The physical component includes all aspects of currents, mixing, light penetration, stratification, and ice; in connection with this, one often speaks of the physical pump bringing denser surface water down to deeper layers and advecting bottom water up to the surface layers. The chemical component treats various aspects of chemical reactions and involves equations based on chemical reactions and stoichiometry. For example, the carbon cycle involves both CO_2 solubility and chemical reactions. One can speak of a solubility pump as, for example, CO_2 dissolves more easily in cold water. The biological pump results from phytoplankton taking up CO_2 and nutrients from the surface layer of the ocean, where photosynthesis takes place, and converting it into organic carbon. By sinking and subsequent mineralization of this carbon, the organic carbon is transformed back into CO_2 and nutrients.

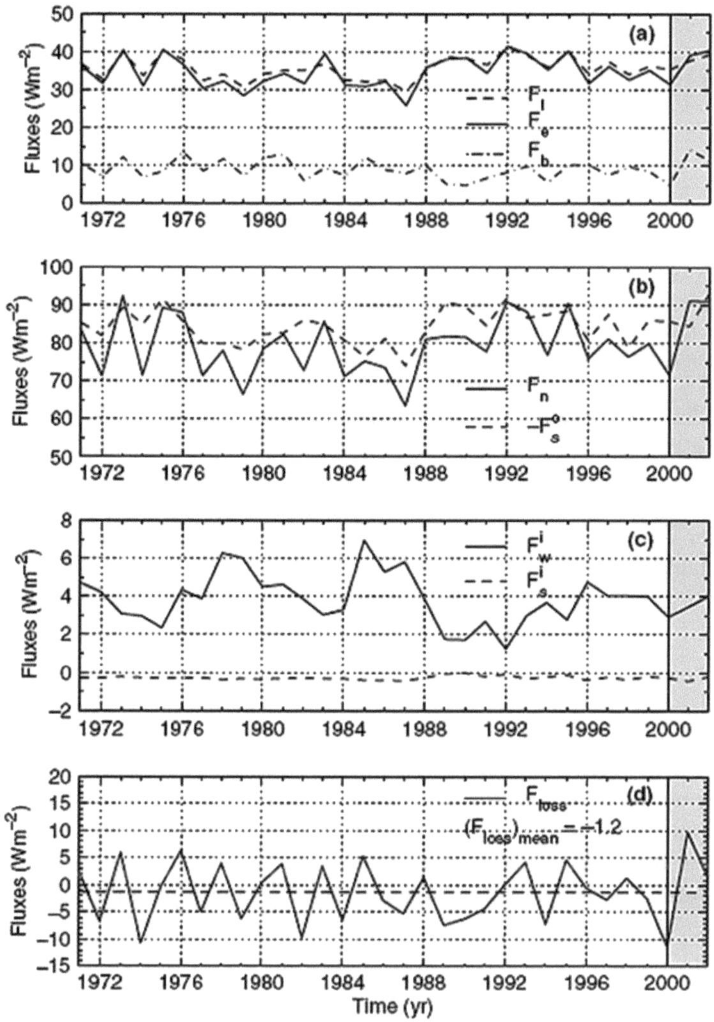

Fig. 3.12 Annual means of: Sensible heat (Fh), latent heat (Fe), net long-wave radiation (Fl), net heat flux (Fn = Fh + Fe + Fl), sun radiation to the open water surface (Fso), sun radiation through ice (Fsi), heat flow from water to ice (Fwi), and net Baltic Sea heat loss, i.e., Floss = (1 − Ai) (Fso + Fh + Fe + Fl) + Ai(Fsi + Fwi), where Ai is the ice concentration (Omstedt and Nohr 2004)

The physical, chemical, and biological processes actively interact and involve many interesting modelling aspects, of which also knowledge about human activity is needed (Fig. 3.13). Reliable datasets of oceanic, atmospheric and biochemical variables, as well as atmospheric deposition and river loads are needed.

Biogeochemical studies often start from simple stoichiometrically linked water, salt, and nutrient budgets in which, for example, carbon (*C*), nitrogen (*N*), and phosphorus (*P*) are considered. The first step is to establish the water and salt

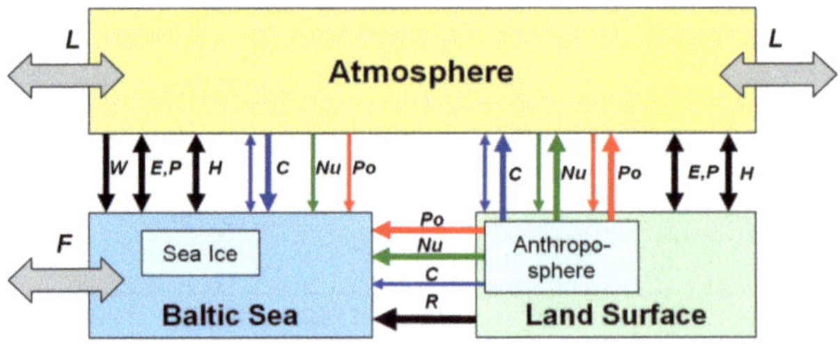

Fig. 3.13 A sketch of the extended BALTEX Phase II research; L denotes lateral exchange with the atmosphere outside the region, W wind stress, E evaporation rate, P precipitation rate, H heat and energy fluxes, R river runoff, F inflows and outflows through the entrance area, C carbon fluxes, Nu nutrient fluxes, and Po pollutant fluxes

balances and calculate the in- and outflows. It should be noted that salinity is a conservative property suitable for establishing the exchange of water between the system and adjacent seas. Total alkalinity can also be regarded as a conservative property and give additional information on, for example, the total alkalinity in the river water entering the system (Hjalmarsson et al. 2008). The next step is to establish budgets for non-conservative materials (Gordon et al. 1996),

$$\frac{dV^{\phi}C}{dt} = {}^{\phi}C_iQ_i - {}^{\phi}CQ_o + \Delta^{\phi}C,$$

where ${}^{\phi}C$ is the concentration of the non-conservative material (e.g., in mol/kg) and $\Delta^{\phi}C$ is the sum of all non-conservative fluxes. The equation allows the quantification of the net non-conservative reaction of material in the system.

Stoichiometric relationships and chemical reaction formulas are used to describe the proportions of the compounds involved in chemical reactions. For example, the carbon cycle can be related to organic material due to primary production, as follows:

$$\begin{aligned} &aCO_2 + bNH_3^- + cH_3PO_4 + bH^+ + (a+b)H_2O \\ &\Leftrightarrow (CH_2)_a(NH_3)_b(H_3PO_4)_c + (a+2b)O_2 \end{aligned}$$

where $(a{:}b{:}c) = (106{:}16{:}1)$ are the standard *Redfield* values. The expression illustrates how carbon dioxide, water, nutrients and oxygen may form plankton in its simplest form $((CH_2)_a(NH_3)_b(H_3PO_4)_c)$. For a discussion about the Redfield concept and marine biogeochemistry, see Sarmiento and Gruber (2006).

Figure 3.14 depicts some processes related to eutrophication of the Baltic Sea (Savchuk and Wulff 2007). The importance of oxygen dynamics for the phosphate dynamics is depicted in Fig. 3.15. At low oxygen concentrations, the sediments start to leak phosphate. From observations, Conley et al. (2002) demonstrated that the annual change in dissolved inorganic phosphate was positively correlated with the area of sea bottom covered with hypoxic water (water with oxygen concentrations less than 2 mL/L). The figure also indicates that, in oxic water, the sediments may

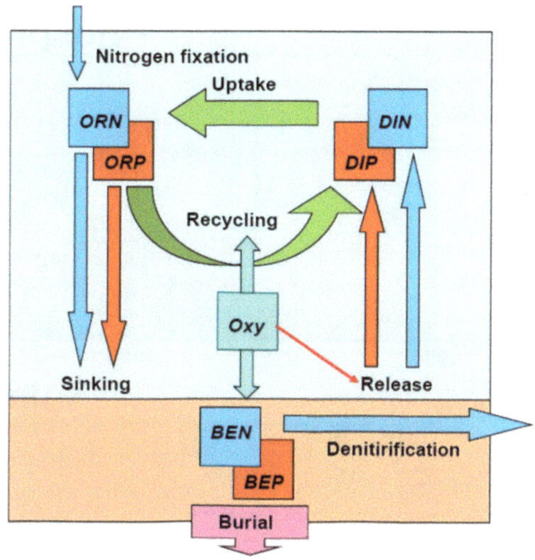

Fig. 3.14 Processes related to modelling the eutrophication in the Baltic Sea based on the modelling approach of Savchuk and Wulff (2007). The organic nitrogen and phosphate parts are denoted by ORN and ORP, the inorganic parts by DIN and DIP, and the benthic parts by BEN and BEP, respectively

bind phosphate. However, the nitrate dynamics are also influenced, and if the water volume turns anoxic this will result in a nitrogen sink, as the NO_3 in the water is reduced to N_2. Any modelling of eutrophication must therefore consider how low oxygen concentrations influence nutrient dynamics.

The presence of many strongly independent variables and the frequent absence of clear basic laws make marine ecosystems difficult to describe quantitatively. There are many kinds of marine organisms that transform nutrients and inorganic carbon into organic material. The primary production of the world's oceans is based on diatoms, dinoflagellates, coccolithophorids, silicoflagellates, and blue–green and other bacteria, all of which need to be understood quantitatively. The regional and species distributions vary considerably, as do light and nutrient limitations. Ojaveer and Kalejs (2008) argued that the Baltic Sea should be divided into ten ecosystem-based regions based on several considerations, bottom topography and currents being two major ones.

3.13 Acid–Base (pH) Balance

While water and heat balances are at the heart of the climate system, the carbon cycle lies at the heart of biogeochemical modelling. Strong efforts are being made to connect climate and biogeochemical models and to include them in Earth System Modelling; the development of these models is a major component of BALTEX Phase II.

The acid–base (pH) balance is somewhat more complicated than the heat or water balance, due to the buffer effect. The balance is controlled by the amount of protons (H^+ ions) added to (donors) or subtracted from (acceptors) the balance. The protons, however, also interact with the buffer system, which can be understood as the total

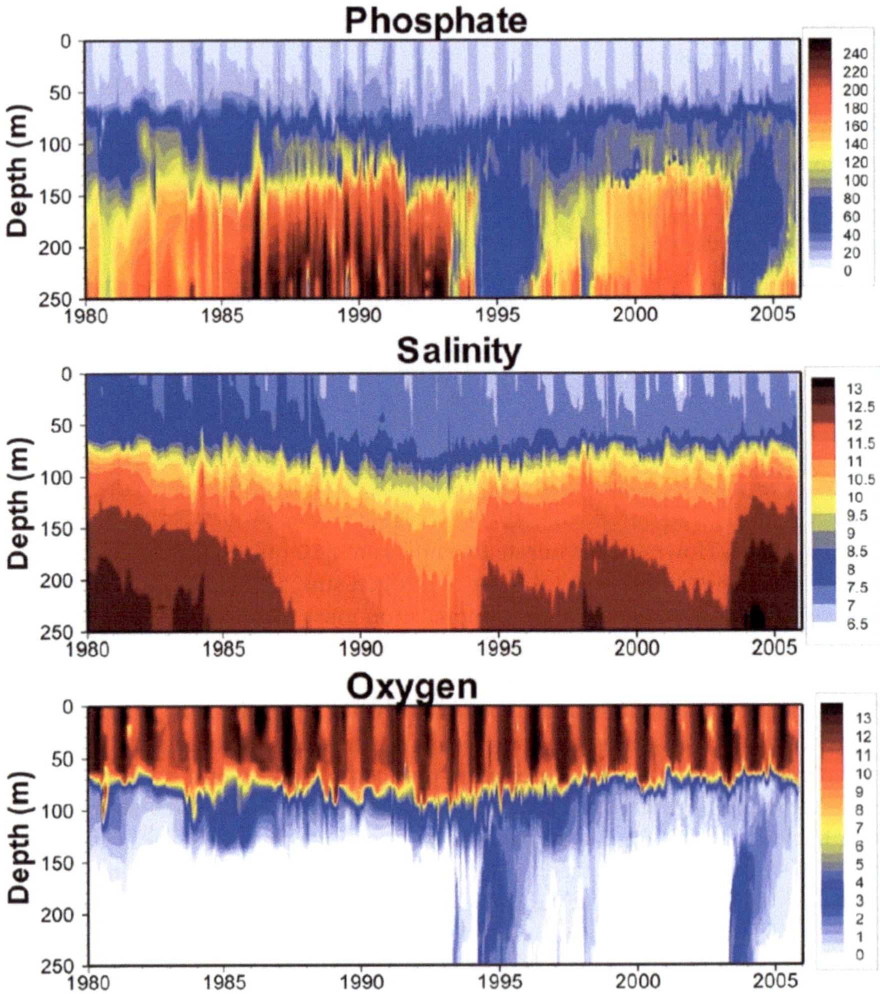

Fig. 3.15 Measurements from the Gotland Deep in the Baltic Sea (From Stigebrandt and Gustafsson 2007)

alkalinity (A_T). The total alkalinity is a major component of the ocean carbon system, and can be defined as "the excess of proton acceptors over proton donors" (Dickson 1981). The major proton acceptors in seawater are hydrogen carbonate (HCO_3^-, 95.5%), carbonate (CO_3^{2-}, 4%) and borate ($B(OH)_4^-$, 0.5%) ions, whereas hydrogen ions (H^+) and hydrogen sulphate ions (HSO_4^-) act as proton donors. A_T is impacted by, for example, dissolution of limestone through the increase of carbonate ions. This addition of carbonate ions strengthens the buffering capacity (A_T) and increases the pH. The addition of weak acids such as CO_2, which does not affect A_T, lowers the pH. However, the effect is small since most of the added CO_2 reacts with CO_3^{2-}(buffer reaction: $CO_2 + CO_3^{2-} + 2H_2O \leftrightarrow 2HCO_3^-$) and only a small fraction of the CO_2

adds hydrogen ions to the system by dissociation of H_2CO_3 to HCO_3^- and H^+. The problem is that when enough carbonic acid (or any other acid) has been added, there are not enough carbonate and bicarbonate ions left, and with further additions, pH decreases more rapidly. This has been observed in freshwater systems exposed to airborne sulphuric acid. Freshwater systems have in general much weaker buffering capacities than marine systems (Fig. 3.16).

To estimate the pH of the water, one must consider the concentration of dissolved inorganic carbon, the major chemical component of the ocean carbon cycle. This cycle includes chemical reactions associated with the dissolving of carbon dioxide from the atmosphere in the surface water, where it undergoes rapid chemical reactions, only a small fraction remaining as carbon dioxide. Limestone, however, may also dissolve, generating carbon ions. The reactions include the following components of the inorganic carbon cycle: carbon dioxide (CO_2), carbonic acid (H_2CO_{3aq}), bicarbonate (HCO_{3aq}^-), and carbonate (CO_{3aq}^{2-}), the sum of which is referred to as the dissolved inorganic carbon (C_T). The state variables are C_T and A_T, defined as follows:

$$C_T = [CO_2] + [HCO_3^-] + [CO_3^{2-}]$$
$$A_T \approx [HCO_3^-] + 2[CO_3^{2-}]$$

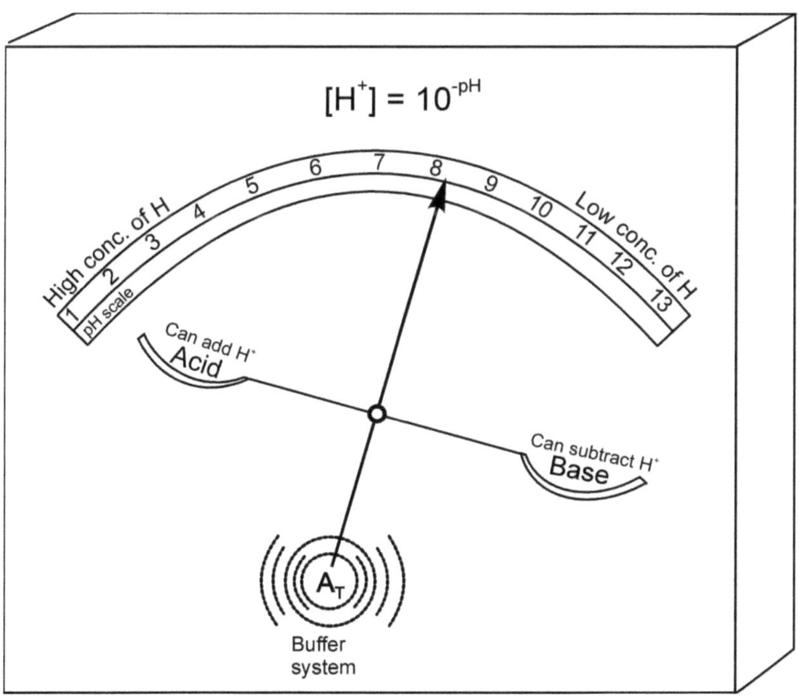

Fig. 3.16 The acid–base (pH) balance depicted as a balance between the concentration of proton donors and proton acceptors, damped by the buffer system represented by the total alkalinity (AT)

If the total inorganic carbon and total alkalinity are known, we can derive a simplified relationship for the pH by ignoring the presence of boric acid. This simplified analytical relationship is as

$$[H^+] \approx \frac{K_2(2C_T - A_T)}{A_T - C_T},$$

where K_2 is a solubility constant, which is temperature and salinity dependent. The mathematical behaviour the equation indicates that pH ($[H^+] = 10^{-pH}$) is mainly dependent on the difference between total alkalinity and total inorganic carbon. In addition, small differences between these parameters can drastically reduce the pH. Figure 3.17 presents typical variations of some of the parameters.

Fossil fuel burning is likely to have both a direct and indirect effect through increased CO_2 levels, altering seawater pH and total alkalinity as well as through changes in river chemistry. This may severely threaten some species in the Baltic Sea, particularly in the Northern Baltic where the buffet system is much weaker than in the rest of the Baltic Sea (Omstedt et al. 2010).

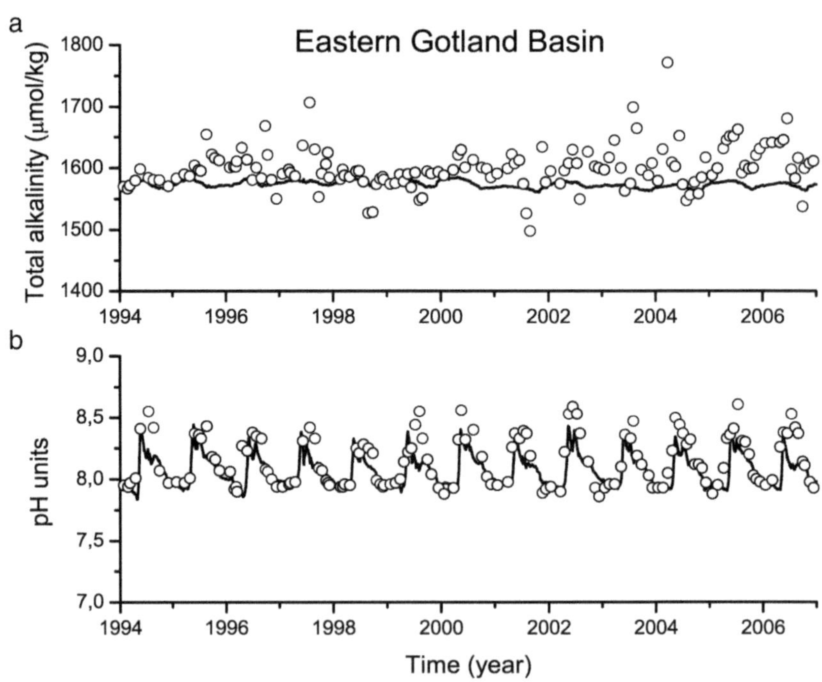

Fig. 3.17 Typical variations of total alkalinity and pH in the surface water of the Baltic Sea (From Omstedt et al. 2009). The *circles* indicate observations and the *fully drawn lines* model simulations

3.14 Some Comments Related to Climate Change

The observation of climate variables (e.g. temperature, precipitation, etc.) constitutes the basis of our understanding of climate and climate change. However, what we subjectively see from data depends greatly on what we are expecting and on our scientific training (BACC 2008). Figure 3.18 presents a time series as both original and normalized data. In the figure, the same time series is interpreted in three different ways. The original dataset illustrates how the maximum annual ice extent (MIB) in the Baltic Sea has varied from 1,720 up to now. Obviously, the data are very noisy. When scientists communicate their results, to decision makers or the general public, the message is often given a political implication. Presenting a trend, the message implies that society must do something. When we present oscillation, the message is that the society does not need to take any action. Finally, when we are speaking about regime shifts, the message to society is that we found something that we will not be able to change. How scientists address their message to the public implies strong signals for different actions. Hence, there is a need for scientist to improve in making sense of science (e.g. Pielke 2007, von Storch, this book).

The MIB time series is a key data source for understanding the Baltic Sea climate, and has recently been extended back to AD 1500 (Hansson and Omstedt 2008). MIB is only an estimated number for each year, but represents an integrated measure of the severity of the ice winter. Proxy information about climate change based on MIB, NAO (North Atlantic Oscillation), and Stockholm sea levels gives us the first simplified view of climate change and variability. Together with other long-term datasets Eriksson et al. (2007) characterize the last 500 years of climate, illustrating an event type conceptual behaviour of climate change with 15 time periods (Fig. 3.19).

Climate and weather show a stochastic and periodic behaviour, and it is necessary to look into the past for long time periods to understand this behaviour. A rule of thumb is that to understand the coming 100 years, we must understand at least 1,000-year periods back in time, taking us back to AD 1,000. From the record of the past 500 years, it seems as though the change from the warm 1730s to the extremely cold year 1740 spans the climate variability range in the Baltic Sea region for the last 500 years (Jones and Briffa 2006). Another interesting consideration is from studies of the Greenland Ice and the Baltic Sea sediment records, Kotov and Harff (2006) have identified periodicities of 900, 500 and 400 years periods which they claim reflects climate processes relevant at least in the North-Atlantic region. We must thus consider both periodical or non-chaotic and chaotic aspects of our time series.

Figure 3.20 indicates the importance of long-term datasets. The Baltic Sea mean salinity exhibits large variations over long time scales (Winsor et al. 2001, 2003), so statements of systematic increases or decreases may be derived when limited segments of many years are considered. However, the time series extending across the entire last century indicates that speaking of a long-term trend makes little sense. Figure 3.21 presents the Stockholm air temperature, indicating a warmer climate since approximately 1880, which has been identified as marking the end of the "Little Ice Age" in our region (Omstedt and Chen 2001).

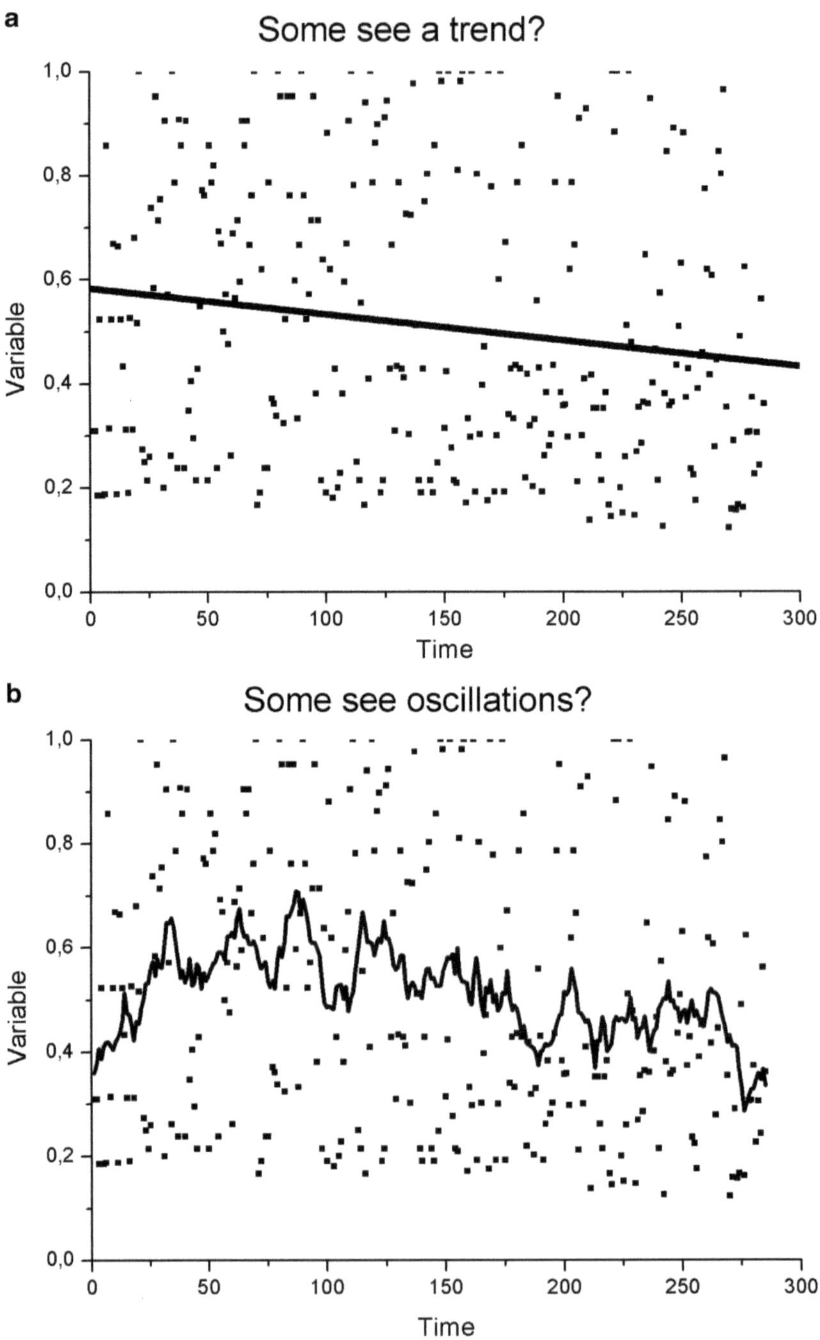

Fig. 3.18 (continued)

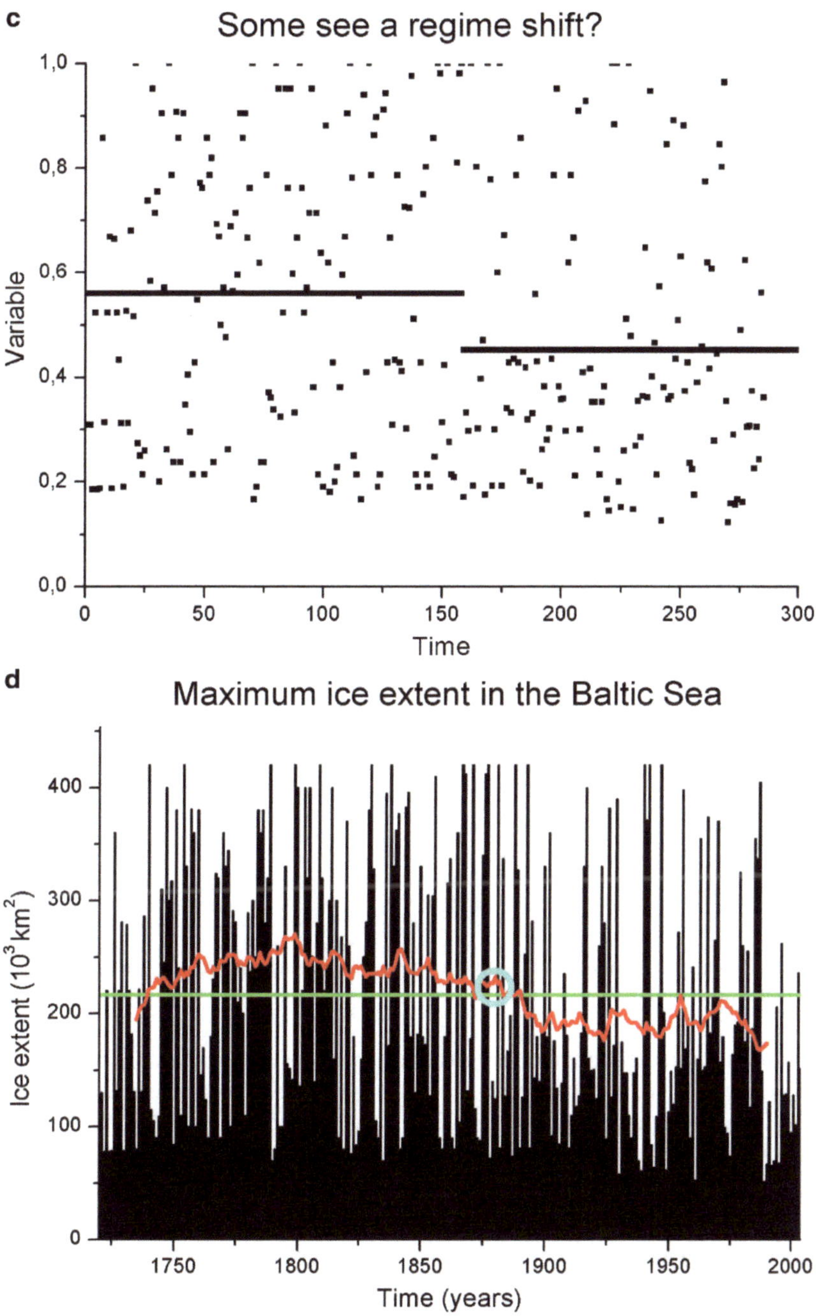

Fig. 3.18 Climate change can be detected by trends, oscillations and jumps or regime shifts. In this figure, the same data are used and normalized (**a–c**). The original data set is illustrated in (**d**) (Redrawn from BACC 2008)

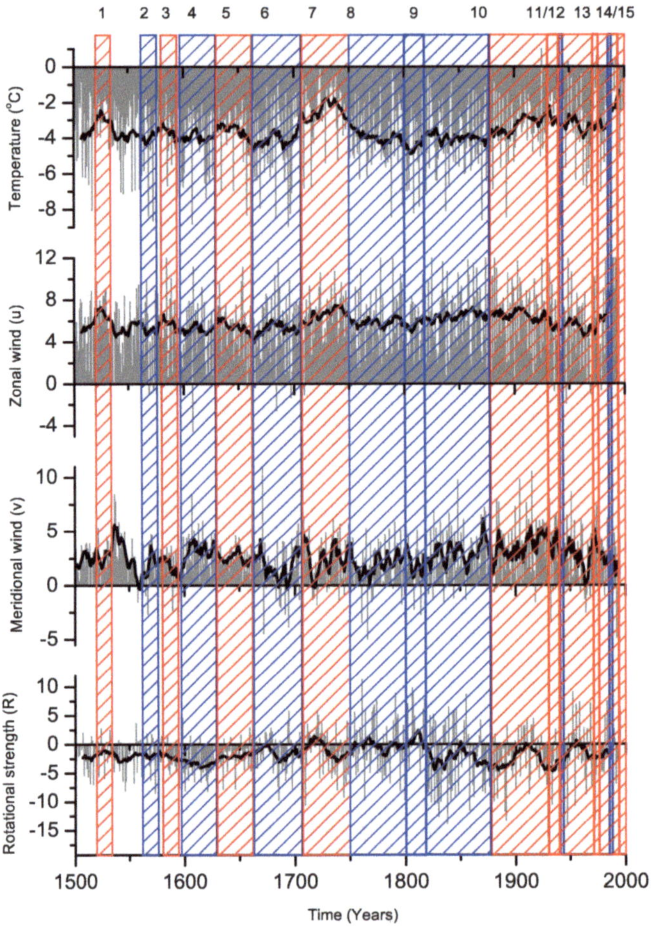

Fig. 3.19 Characterizing the 1500–2000 climate period into 15 different warm and cold periods (From Eriksson et al. 2007)

Important climate change questions are: How sensitive is the climate to changes in solar irradiance, atmospheric aerosols, greenhouse gases (including water vapour), and other climate forcing? How large is the natural climate variability? Based on proxy data, the climate forcing factors have been estimated by several authors (e.g., Crowley 2000). Figure 3.22 depicts the change in radiative forcing over the last 1,000 years. The forcing sensitivity between surface temperature (ΔT_s) and radiation changes (ΔF) can be studied by considering the net irradiance at the tropopause, which is believed to be a good indicator of the global mean temperature (globally and annual averaged), defined as:

$$\Delta T_s / \Delta F = \lambda$$

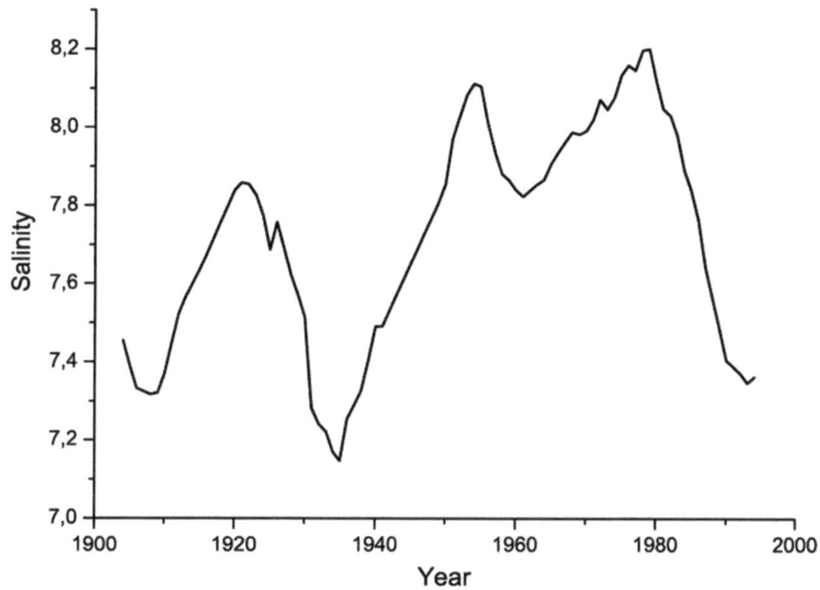

Fig. 3.20 The Baltic Sea mean (annually, horizontally and vertically averaged) salinity (Redrawn from Winsor et al. 2001, 2003)

(IPCC 2001, p. 354)where λ is a constant that, in simplified radiation–convection models, is typically approximately 0.5 K/Wm^{-2}. A change in the radiation balance of 2 Wm^{-2} thus corresponds to a global temperature increase of 1 K.

Currently, the dominant concern is the effects of increased greenhouse gases and aerosols (IPCC 2007). The increased CO_2 content of the atmosphere increases the greenhouse effect and therefore warming (the first climate change aspect). How strongly atmospheric CO_2 will increase the atmospheric temperature is under discussion, but it is generally believed to primarily influence the global radiation balance, but also cloud dynamics. The second aspect of increasing atmospheric CO_2 is its effects on the acid–base balance in the ocean. Increased atmospheric CO_2 will reduce the ocean pH, which may threaten marine ecosystems (see Brander and Havenhand this book). A third aspect of increasing atmospheric CO_2 is its effects on land vegetation and increased biological production in regions where water is not a limiting factor. Increased biological production may lead to increasing levels of dissolved organic carbon entering the coastal seas through rivers. Through mineralization processes, this could become an additional source of acid material in these seas.

Long time series are scarce before the eighteenth century, so we should accept that, going back several hundred years, we will probably underestimate variability in both space and time (Storch et al. 2004). Different feedback mechanisms may increase and/or decrease climate change and cause regional differences. Not surprisingly, it is not easy to attribute the present climate change to one single cause, and it is even more difficult to attribute causes on the regional scale (BACC 2008).

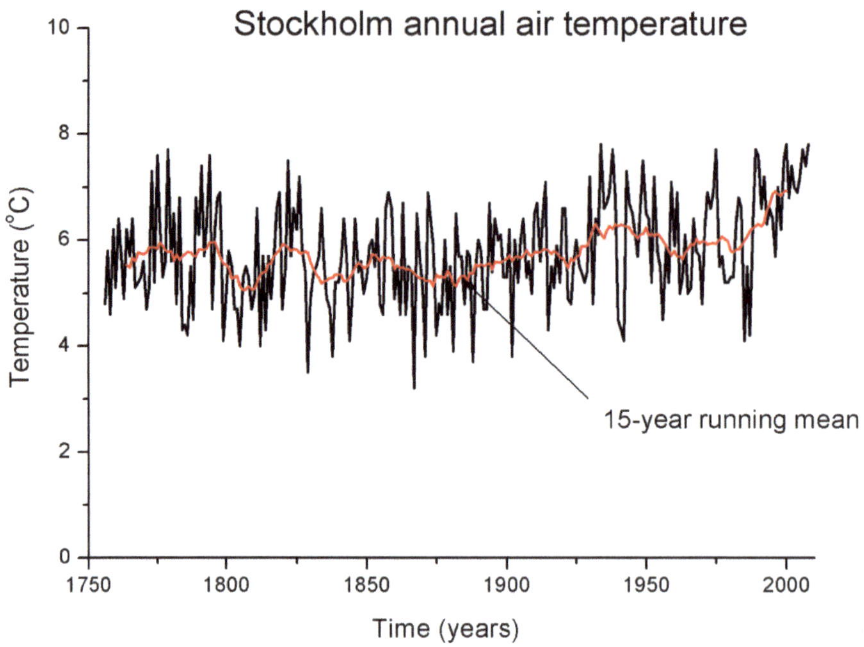

Fig. 3.21 Stockholm annual air temperature adjusted data for the city effect according to Moberg et al. (2002)

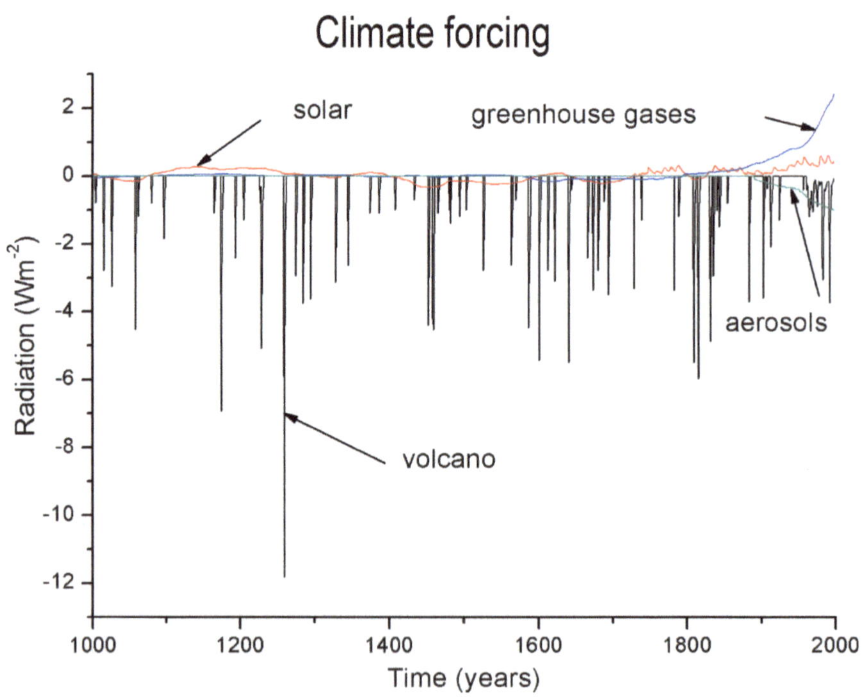

Fig. 3.22 Climate forcing estimations (Redrawn from Crowley 2000)

To attribute the different anthropogenic effects in combination to natural variability is a major research area in the Baltic Basin for the coming years.

3.15 Suggested Further Reading

Some interesting new books are now available which treat Baltic Sea environmental issues related to oceanography and climate change. Omstedt (2011) recently published a book on process oriented modelling of lakes and coastal seas with several examples and exercises from the Baltic Sea. The book by Leppäranta and Myrberg (2009) treats the physical oceanography of the Baltic Sea including sea ice. Feistel et al. (2008) presented the state and evaluation of the Baltic Sea from 1952 to 2005 together with a CD-ROM including Baltic Sea data. An assessment of the Baltic Basin climate change including the Baltic Sea and terrestrial ecosystems is available in the BACC book (2008). Fennel and Neumann (2004) introduced marine ecosystem modelling with applications to the Baltic Sea. The book by Wulff et al. (2001) illustrates some important aspects of the Baltic Sea marine system and material balances.

Acknowledgements The lectures for this summer school and this book contribute to the BONUS + project Baltic-C. The Baltic-C funding was from the European Community's Seventh Framework Program (FP/2007–2013) under grant agreement 217246 made with the BONUS programme and from FORMAS. Useful comments on an earlier draft of this chapter were given by Kari Eilola and an anonymous reviewer.

References

van Aken HM (2007) The oceanic thermohaline circulation. An introduction, vol 39, Atmospheric and Oceanographic Sciences Library. Springer, New York, p 326

BACC Author Team (2008) Assessment of climate change for the Baltic sea basin, Regional Climate Studies. Springer, Berlin/London. doi:ISBN: 978-3-540-72785-9

Conley DJ, Humborg C, Rahm L, Savchuk OP, Wulff F (2002) Hypoxia in the Baltic Sea and basin-scale changes in phosphorus biogeochemistry. Environ Sci Technol 36:5315–5320

Crowley TJ (2000) Causes of climate change over the past 1000 years. Science 289(5477): 270–277. doi:DOI: 10.1126/science.289.5477.270, 14 July

Cushman-Roisin B, Becker JM (2011) Introduction to Geophysical Fluid Dynamics: Physical and Numerical Aspects. Second edition. Vol 101 in the International Geophysical Series. Academic Press, Elsevier, Amsterdam, The Netherlands

Dickson AG (1981) An exact definition of total alkalinity and a procedure for the estimation of alkalinity and total inorganic carbon from titration data. Deep-Sea Res 28A:609–623

Eriksson C, Omstedt A, Overland JE, Percival DB, Mofjeld HO (2007) Characterizing the European subarctic winter climate since 1500 using ice, temperature, and atmospheric circulation time series. J Climate 20:5316–5334. doi:10.1175/2007JCLI1461.1

Feistel R, Nausch G, Wasmund N (2008) State and evaluation of the Baltic sea, 1952–2005. A detailed 50-year survey of meteorology and climate, physics, chemistry, biology, and marine environment. Wiley, Hoboken

Fennel W, Neumann T (2004) Introduction to modelling the marine ecosystems, vol 72, Elsevier oceanography series. Elsevier, Amsterdam/Boston

Gill AE (1982) Atmosphere–ocean dynamics, vol 30, International Geophysics Series. Academic Press, New York

Gordon DC Jr, Boudreau PR, Mann KH, Ong J-E, Silvert WL, Smith SV, Wattayakorn G, Wulff FT, Yanagi T (1996) LOICZ biogeochemical modeling guidelines. LOICZ/R&S/95-5, VI +96 pp, LOICZ, TEXEL, the Netherlands

Green MJA (2004) Dynamics of the upper coastal ocean with special reference to the inshore-offshore water exchange. PhD thesis A89, Department of Earth Sciences: Oceanography, University of Gothenburg, Sweden

Green MJA, Liljebladh B, Omstedt A (2006) Physical oceanography and water exchange in the Northern Kvark Strait. Cont Shelf Res 26:721–732. doi:10.1016/j.csr.2006.01.012

Hansson D, Omstedt A (2008) Modelling the Baltic Sea ocean climate on centennial time scales; temperature and sea ice. Clim Dyn 30(7–8):763–778. doi:10.1007/s00382-007-0321-2

Hjalmarsson S, Wesslander K, Anderson LG, Omstedt A, Perttilä M, Mintrop L (2008) Distribution, long-term development and mass balance calculations of total alkalinity in the Baltic Sea. Cont Shelf Res 28(4–5):593–601

IPCC (2001) In: Houghton et al. (eds) Climate change: the physical science basis. Contribution of working Group I to the third assessment report of the intergovernmental panel on climate change. Cambridge University Press, Cambridge, UK, 881 pp

IPCC (2007) In: Solomom S, Qin D, Manning M, Chen Z, Marquis M, Averyt KB, Tignor M, Millers HL (eds) Climate change: the physical science basis. Contribution of working group 1 to the fourth assessment report of the intergovernmental panel on climate change. Cambridge University Press, Cambridge, UK, 996 pp

Jones PD, Briffa KR (2006) Unusual climate in northwest Europe during the period 1730 to 1745 based on instrumental and documentary data. Clim Chang 79:361–379. doi:10.1007/s10584-006-9078-6

Kotov S, Harff J (2006) A comparison of Greenland Ice and Baltic Sea Sediment Record – a contribution to climate change analysis. Math Geol 38(6), DOI: 10.1007/s11004-006-9047-7

Leppäranta M, Myrberg K (2009) Physical oceanography of the Baltic sea. Springer Praxis, Chichester

Mattsson J (1996) Some comments on the barotropic flow through the Danish Straits and the division of the flow between the Belt and the Öresund. Tellus 48:456–471

Moberg A, Bergström H, Krigsman JR, Svanered O (2002) Daily air temperature and pressure series for Stockholm (1756–1998). Clim Change 53:171–212

Ojaveer E, Kalejs M (2008) On ecosystem-based regions in the Baltic Sea. J Mar Syst 74:672–685

Omstedt A (2011) Guide to process based modelling of lakes and coastal seas. Springer-Praxis Books in Geophysical Sciences. Springer, Berlin/Heidelberg. doi:10.1007/978-3-642-17728-6

Omstedt A, Rutgersson A (2000) Closing the water and heat cycles of the Baltic Sea. Meteorol Z 9:57–64

Omstedt A, Chen D (2001) Influence of atmospheric circulation on the maximum ice extent in the Baltic sea. J Geophys Res 106(C3):4493–4500

Omstedt A, Axell L (2003) Modeling the variations of salinity and temperature in the large Gulfs of the Baltic sea. Cont Shelf Res 23:265–294. doi:10.1016/S0278-4343(02)00207-8

Omstedt A, Nohr C (2004) Calculating the water and heat balances of the Baltic Sea using ocean modelling and available meteorological, hydrological and ocean data. Tellus 56A:400–414

Omstedt A, Elken J, Lehmann A, Piechura J (2004) Knowledge of the Baltic Sea physics gained during the BALTEX and related programmes. Prog Oceanogr 63:1–28. doi:10.1016/j.pocean.2004.09.001

Omstedt A, Hansson D (2006a) The Baltic Sea ocean climate system memory and response to changes in the water and heat balance components. Cont Shelf Res 26:236–251. doi:10.1016/j.csr.2005.11.003

Omstedt A, Hansson D (2006b) Erratum to: 'The Baltic Sea ocean climate system memory and response to changes in the water and heat balance components'. Cont Shelf Res 26:1685–1687. doi:10.1016/j.csr.2006.05.011
Omstedt A, Gustafsson E, Wesslander K (2009) Modelling the uptake and release of carbon dioxide in the Baltic Sea surface water. Cont Shelf Res 29:870–885. doi:10.1016/j.csr.2009.01.006
Omstedt A, Edman M, Anderson LG, Laudon H (2010) Factors influencing the acid–base (pH) balance in the Baltic Sea: a sensitivity analysis. Tellus 62B:280–295. doi:10.1111/j.1600-0889.2010.00463.x
Piechura J, Beszczynska-Möller A (2004) Inflow waters in the deep regions of the southern Baltic Sea-transport and transformations. Oceanologia 46(1):113–141
Pielke RA Sr (2003) Heat storage within the earth system. Br Am Meteorol Soc 84:331–335
Pielke RA Jr (2007) The honest broker. Making sense of science in policy and politics. Cambridge University Press, Cambridge
Sarmiento JL, Gruber N (2006) Ocean biogeochemical dynamics. Princeton University Press, Princeton
Savchuk OP, Wulff F (2007) Modeling the Baltic Sea eutrophication in a decision supporting system. Ambio 36(2–3):141–148
Stigebrandt A (2001) Physical oceanography of the Baltic Sea. Chapter 2 In: Wulff F, Rahm L, Larsson P (eds) System analysis of the Baltic Sea, vol 148, Ecological Studies. Springer, Berlin
Stigebrandt A, Gustafsson BG (2007) Improvement of Baltic proper water quality using large-scale ecological engineering. Ambio XXXVI(2–3):280–286. doi:doi:10.1029/2006JG000304, G02023
von Storch H, Zorita E, Jones J, Dimitriev Y, González-Rouco F, Tett SFB (2004) Reconstructing past climate from noisy data. Science 306(5696):679–682. doi:10.1126/science.1096109
Versteeg HK, Malalasekera W (1995) An introduction to computational fluid dynamics. The finite volume method. Addison Wesley Longman, England
Winsor P, Rodhe J, Omstedt A (2001) Baltic Sea ocean climate: an analysis of 100 yr of hydrographic data with focus on the freshwater budget. Clim Res 18:5–15
Winsor P, Rodhe J, Omstedt A (2003) Erratum: Baltic Sea ocean climate: an analysis of 100 yr of hydrographical data with focus on the freshwater budget. Clim Res 18:5–15, 2001. Clim Res 25:183
Wulff F, Rahm L, Larsson P (2001) System analysis of the Baltic Sea, vol 148, Ecological Studies. Springer, Berlin

Chapter 4
A Basic Introduction to Climate Modeling and Its Uncertainties

Eduardo Zorita

4.1 Climate Drivers, Feedbacks, Forced and Internal Variability

The main tool for estimating possible climate changes associated with the increase of carbon dioxide (CO_2) and methane (CH_4) in the atmosphere are numerical climate models. In theory, the climatic influence of an increasing concentration of CO_2 in the atmosphere could be calculated in a relatively simple manner, based on the known laws of physics and chemistry. However, the large number of components of the climate system, and the non-linear (chaotic) interactions between them makes this endeavor impossible. Modern computer climate models try to represent the known physical laws in an approximate way on a grid spanning the Earth's surface, the three-dimensional atmosphere and the three-dimensional ocean. The components of the climate system are thus divided into small cells, each spanning a few degrees of latitude and longitude. These cells interact with each other at well defined time steps, exchanging energy, momentum, and matter. An illustrative, albeit only qualitatively accurate analogy would be a three-dimensional chess board, each cell of which is occupied by a person. At the whistle of a referee (roughly every 15 min), all persons simultaneously exchange energy and matter with the person in the neighbouring cell. The rules each person obeys are a representation of the true physical laws. It is therefore important to keep in mind that numerical climate models are only an approximation of reality. However, modern models have reached a high degree of sophistication, and they can realistically represent the global climate in many aspects. In other aspects they still grapple with difficult problems.

E. Zorita (✉)
Institute of Coastal Research, Helmholtz-Zentrum Geesthacht, Max-Planck-Strasse 1, Geesthacht, Germany
e-mail: eduardo.zorita@hzg.de

M. Reckermann et al. (eds.), *Climate Impacts on the Baltic Sea: From Science to Policy*, Springer Earth System Sciences, DOI 10.1007/978-3-642-25728-5_4,

An important concept is that of climate forcings. A climate forcing, or driver, is a factor which provides or modulates the exchange of energy with the extraterrestrial space but which it is not influenced itself by the internal dynamics of the climate. A clear example of an external driver is the sun, which ultimately provides virtually all the energy to the different components of the climate system. The solar energy reaching the upper layers of the atmosphere (with an annual mean value of currently about 1,365 W/m^2) may vary but it is completely independent of any possible changes in the Earth's climate. In the case of the sun its classification as external driver is quite clear, but for other drivers a little more attention is required. For instance, volcanic eruptions are also considered as external climate forcings. The sulfuric gases ejected in a volcanic eruption form aerosols, part of which reach the stratosphere – the stratified and stable layer of the atmosphere above the turbulent and well mixed lower layer, the troposphere. If the eruptions are powerful enough, ejecta may remain in the stratosphere for a few years until they slowly fall into the troposphere again and are washed out from the atmosphere by rain. In the stratosphere, volcanic derived aerosols modify the amount of solar radiation entering the troposphere, thus modifying the amount of energy entering the climate system. However, the distribution of volcanic aerosols is modulated by the winds in the stratosphere so that strictly speaking, they are not a completely external factor. This effect on the radiative properties of aerosols is however small, and for practical purposes volcanic aerosols are also considered as external forcings. The same can be said of the atmospheric concentrations of greenhouse gases, CO_2, presently at about 390 ppm, and rapidly increasing, and CH_4 (presently at 1,700 ppb with a much more moderate growth), and other trace gases. Greenhouse gases modulate the interchange of energy between atmospheric layers. Ultimately, they act as a valve regulating the amount of infrared energy that can be radiated back to space from each atmospheric layer. Their concentration in the atmosphere is also influenced by atmospheric dynamics. Furthermore, on long time scales (over centuries), the atmospheric concentration of CO_2 can change considerably as part of the global carbon cycle. Carbon dioxide can be stored for a long time in carbon reservoirs in the ocean and in the biosphere. Again, for shorter time scales, a useful simplification is to consider CO_2 as a well mixed gas in the troposphere and stratosphere, the concentration of which is either almost constant or increasing along a predetermined path which is independent of climate. In this latter case, the CO_2 concentration would be a climate forcing factor. In the former case, in which it is allowed to vary interactively with other components of the climate system, the corresponding climate driver would be the carbon *emissions* by anthropogenic activities or, at longer time scales, by volcanic eruptions. Another important external climate driver is changes in land use. The different types of land surfaces, forest, crops, tundra, etc. have different reflective properties (to solar radiation) and also different dynamical properties, such as resistance to wind flow, so that they also can modify climate. Strictly speaking, different land use types also feel the influence of climate since the land biosphere reacts to changes in the temperature and precipitation. Again, a useful simplification is to consider that the type of land cover is either constant or changing along a prescribed path, i.e. not interactively changing with the climate itself. In this sense, it can be considered an external climate forcing factor. If, on the other hand, the evolution of land cover is interactively changing according to the effect of other climate components, it cannot be considered external forcing any more. From these examples it should be clear that what constitutes external forcing depends on the system under study is: external forcings are processes that may influence our system but themselves are not influenced by it. An external forcing factor may thus be considered as an internal process when the studied system is extended.

The four factors just mentioned are the most important external drivers of the climate system. If these factors change, the mean temperature of the Earth will also change, as a result of the changing energy balance. However, the modified energy balance may induce further changes in precipitation, cloud cover, etc. These induced changes may in turn affect the exchange of energy between the Earth and the extraterrestrial space. For instance, an increase in cloud cover, in response to external forcing, may then reflect more solar radiation back to space, damping the forcing affect on the mean climate state. Internal mechanisms which evolve in interaction with other climate variables but also affect the exchange of energy of the Earth in this way are called feedbacks. For instance, an important question is what the magnitude of the climate response to changes in the atmospheric concentrations of CO_2, is, the so-called climate sensitivity, which will be defined more precisely later. If the solar output increases by 1 per mill, the mean temperature of the earth will tend to increase and, together with temperature, many other factors will also change, among them cloud cover which in turn modulates the amount of radiation reaching the surface. Finally, when a new equilibrium is reached, the total temperature response experienced after an increase of 1% in the incoming solar energy will also be a function of all possible feedbacks present in the climate system. The magnitude of the equilibrium temperature response when the concentration of CO_2 in the atmosphere is doubled turns out to be more dependent on the feedbacks than on the direct perturbation caused by CO_2.

Another source of climate variation is generated endogenously by the chaotic nature of the climate system, and not by the effect of external forcings. It is obvious for everyone that temperature and precipitation are different every day, every week, every year, and perhaps not so obvious, but also every decade and every century. These variations occur even when external climate drivers were strictly constant. The trajectory of these internal climate variations depend on a number of factors, for instance on the current state of the system (the so called initial conditions). Although theoretically it is possible to calculate the trajectory of the climate system, if the positions and velocities of all fluids and particles are known, it is not possible in practical terms. Any errors in the determination of those initial conditions will grow with time, and after a certain predictability limit, predictions and the reality diverge. This is familiar to all of us, since for instance we known that weather predictions are not useful for forecast times of about 1 week. Students should be aware that this predictability limit refers only to the internal climate variations, and not to the externally forced variations. For instance, it is difficult to predict whether temperatures 10 days ahead in a certain location will be warmer or cooler than today's temperatures. But it is very easy to predict that temperatures on Bornholm on January 15, 2010 will in all probability be colder than temperatures on July 15, 2010 at the same location.

The observed climate variations are therefore a mixture of two contributions: one externally driven when the external factors change, and one endogenously generated. Climate models contain both types of climate variations as well: when the solar output increases in a climate model, the simulated temperature will increase, following a certain path. However, if we note the trajectory of a simulated climate until the equilibrium has been reached, we observe that this trajectory depends on the initial conditions from which the simulation started. Another simulation with slightly different initial conditions will follow a different path, but in the end the simulation will randomly vary around the same mean state as the first simulation. The change in the mean state has been externally forced, whereas the fluctuating path between both initial and final states is caused by internal variations. The existence of these two types of variations has consequences for the estimation of future climate change, since in general, different simulations for the twenty-first century with the same model, driven by the same atmospheric CO_2 change, will provide slightly different results. In general, at regional scales, the differences among the members of such ensembles of simulations are non-negligible.

4.2 Global and Regional Climate Models

As mentioned before, climate models represent the Earth's climate on a gridded mesh. It would be desirable that this mesh has a very fine spacing between cells since this would increase the accuracy of the representation of the theoretical laws governing the climate system. The regional particularities which modulate the regional and local climate will also be more realistically represented on a finer mesh. However, the grid cell size is limited by the computing power. Halving the grid cell size increases the time needed to carry out the same computation by a factor of 8, since the number of grid cells has quadrupled and the time-steps at which the calculations are carried out must also be halved to limit the growth of errors due to the unavoidable approximate character of the computations. The growth of the required computing power imposes a hard limit on the spatial resolution of climate models as the grid cell size decreases. Currently, this resolution lies roughly between one and three geographical degrees, so that a region like the Baltic Sea area would be represented by about 20 grid cells. This means that the whole climate of the Baltic area would have to be condensed to temperature, precipitation, etc. in about 20 regions. In the modeled world, the climate would be the same within each of these regions. This resolution is too coarse to realistically represent the land topography, the coast lines and exchange between the North Sea and the Baltic Sea. In current climate models, for instance, this pathway is either closed or represented by one (the minimum) or two grid cells, which a width of at least 200 km, but more commonly 400 km.

Regional climate models can be used to ameliorate this problem, as they are able to represent the regional climate much better (see Rummukainen 2010, for a recent review of regional models). The regional models have a much finer resolution than global models (10–50 km), and can be simply considered as a region within a global climate model in which more grid cells were located. In practice, the global climate model and the regional climate model are different entities. The data generated in global simulations are stored and used later to drive the regional climate model at their boundaries, where the coarse global mesh becomes a finer regional mesh. Regional climate models can in principle deliver a better representation not only of the regional characteristics that remain unchanged through time, like topography, but also of small-scale processes like atmospheric convection, storms (cyclones in some regions, for instance the Mediterranean, may develop in a pretty small area), and of course precipitation, especially when orography (e.g. the relief of the Earth's surface) strongly interacts with low-pressure systems. Also, regional models usually contain a better and more detailed representation of soil processes, and thus evaporation over land and run-off should be much more realistic than in global climate models.

It should be mentioned that another strategy exists to by-pass the scale mismatch between global climate models and the need for projections at regional and local scale. This strategy is based on statistical models instead of physical models, and is termed in the literature as *statistical downscaling* (Wilby et al. 2004). The underlying idea of statistical downscaling is to identify statistical relationships between large-scale climate patterns and local or regional climate variables. An example of this type of relationship in the Baltic Sea area would be the North Atlantic Oscillation as a large-scale atmospheric pattern, and run-off into the Baltic Sea as a regional variable. This relationship is mediated by different physical processes, like increased precipitation, changes in evaporation due to temperature or wind etc., which in the end are embodied in a relatively simple statistical model that takes the North Atlantic Oscillation index as a predictor. The parameters of the statistical model are usually estimated using observations of all involved variables in the twentieth century. The statistical model can be then applied to past or future values of the North Atlantic Oscillation index to estimate changes in Baltic Sea run-off. Since this chapter is devoted to climate modeling, statistical downscaling will not be discussed here in detail – the reader is rather referred to the review by Wilby et al. (2004) and references therein. A few advantages and disadvantage of this approach are just mentioned here. The main advantage is the technical simplicity of statistical downscaling compared to simulations with a regional climate model. Also, statistical downscaling can in principle capture, in a simplified form, processes that could be very difficult to include in a regional climate model. The main drawback is that this approach assumes that the statistical relationship derived from observations can be extrapolated into the future unchanged. This assumption may often be plausible, but it cannot be demonstrated, and sometimes it is obviously not correct. In the example cited above, Baltic regional temperature in the future will be driven not only by the North Atlantic Oscillation but more importantly by direct greenhouse gas forcing. Therefore, some care is needed in the design of the statistical model to include the correct predictors for a given target variable.

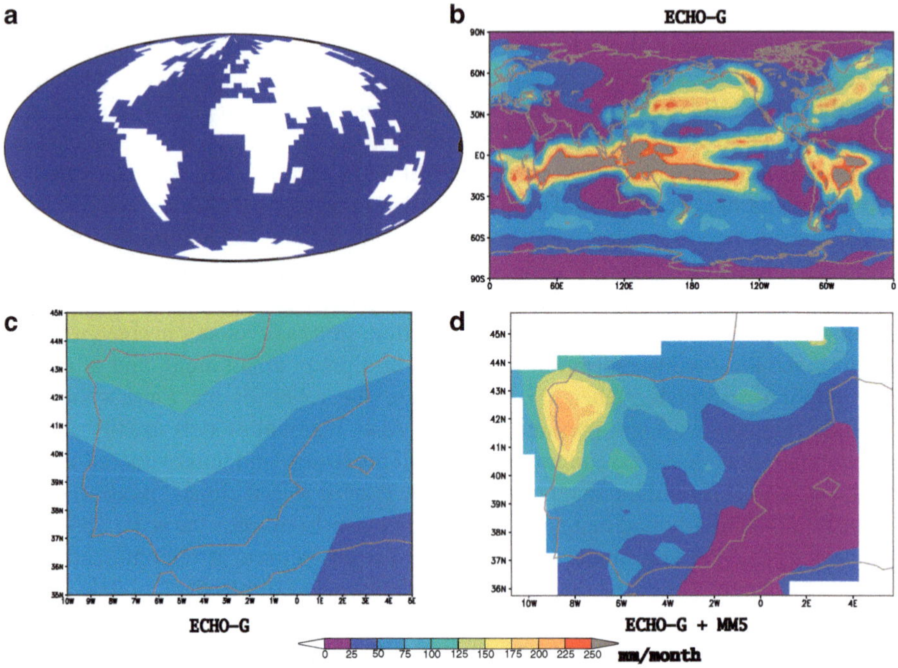

Fig. 4.1 (**a**) (*upper left*) Land-sea mask in a typical global General Circulation Model in which the grid cell size can be recognized; (**b**) (*upper right*) December–February mean precipitation simulated by the model ECHO-G for the present climate; (**c**) (*lower left*) same as (**b**) for the Iberian Peninsula; (**d**) (*lower right*) same as (**c**) but simulated with the regional climate model MM5, driven at the boundaries by the global model ECHO-G. The MM5 simulation region is much wider than depicted in this figure and roughly covers Western Europe, North Africa, the mid-Atlantic Ocean and Western Mediterranean

The reader is also cautioned that the range of validity of statistical methods is usually limited. Nevertheless, statistical downscaling can be a preferable strategy to overcome some problems in regional models.

An example of the limitations of a global climate model and the advantage provided by the use of a regional model are represented in Fig. 4.1. Figure 4.1a displays a typical grid of a global climate model. The global model ECHO-G is part of the model suite used by the Intergovernmental Panel on Climate Change (IPCC) to conduct its projections of future climate. It was developed at the Max-Planck Institute for Meteorology in Hamburg (Germany). It is a global coupled ocean-atmosphere-sea-ice-soil model. It does not contain an interactive model for vegetation nor for the global carbon cycle, and its horizontal resolution is somewhat coarser than in other IPCC models. The stratosphere is also coarsely resolved, which is a common feature among IPCC models. Further technical details about all models mentioned in this chapter can be found in the Appendix of this chapter. The regional model MM5 was developed at the National Center for Atmospheric Research in Boulder (USA), and it has been used by many groups worldwide for weather and environmental prediction. Some groups have also applied this model for longer climate simulations. All global climate models are continuously updated. For instance, the ECHO-G model was superseded by a more modern version ECHAM5-OM, which may include a model for vegetation, carbon cycle, stratosphere chemistry and other processes, and also a newer version of the model MM5 was issued as the model WRF. Usually, newer versions of climate models perform better than older versions, but the student should bear in mind that this is not always the case. Also, an increase in horizontal and vertical resolution does not always bring about an improvement of the simulations of present climate.

The mean annual precipitation simulated by these climate models, under present conditions of solar insolation and atmospheric concentrations of greenhouse gases, is displayed in Fig. 4.1b. This large-scale picture is very realistic, with a precipitation maximum over the Western Pacific and South East Asia. However, when looking at the regional scale, the limitations of the simulation are clearly highlighted in Fig. 4.1c, which shows the mean winter precipitation over the Iberian Peninsula, simulated by the same model. Finally, when the global climate model ECHO-G is used to drive the regional climate model MM5, the improvement in the simulation of precipitation becomes evident (Fig. 4.1d). Over this area, winter precipitation is spatially strongly modulated by the complex topography, and the regional climate model is able to mimic this distribution quite realistically. When comparing data from the climate simulations with present observations, or climate simulations from past periods, with reconstructions of temperature or precipitation based on so-called proxy data, one has to be aware of the limitations of global climate models due to their coarse spatial resolution.

It would be tempting to assume that the resolution problem could be perfectly solved by using regional models with increasingly finer resolution. However, there are some limitations to this approach. Since the physical processes in atmosphere and the ocean display a whole range of spatial scales, from micro-turbulence to the very long planetary waves, no matter how fine the resolution of a model is, there will always be processes that cannot be explicitly *resolved* which nevertheless have to be represented, at least in an indirect way. This is achieved by so called sub-grid *parametrizations*. One obvious example is turbulence occurring at sub-grid scales. It is well known that a turbulent fluid motion is able to disperse fluid tracers, heat, moisture, salt, etc. more effectively than a regular laminar fluid motion. This enhanced capability for diffusion is included in the model by artificially enlarging the diffusion coefficients in the physical equations. Obviously, this is not completely accurate, but it is an ad-hoc heuristic way of including important physical processes that otherwise would be ignored in the models. Another example of parametrization is related to the micro-physical processes that drive the formation of clouds. Clouds display typical spatial scales of a few hundred meters or kilometers, and the condensation and evaporation processes within clouds occur at spatial scales in the range of millimeters and below. It is therefore not possible to represent explicitly the formation of clouds in present climate models. They also have to be parametrized as a heuristic function of the mean humidity, winds, temperature of a specific grid cell and its neighbours in the model. The set of all sub-grid processes that are not explicitly represented in a model is called in general *model parametrizations*.

Parameterizations are different from model to model, since they are both based on experimental results from measuring campaigns and on broad expert assessments. The parametrizations may be responsible for a substantial part of the different behaviours of climate models when simulating present and future climates.

Therefore, as the spatial resolution of a model becomes finer, additional physical processes need to be explicitly modeled and some must be parametrized, but often the parametrizations also have to be modified. For instance, convective processes that would be represented in an averaged (*parametrized*) form in a model with coarse resolution, now would have to be explicitly modeled. Some approximations to the laws of physics which may be valid at coarse resolutions, need to be reformulated at higher resolutions. For instance, at coarse resolution, a simplifying assumption usually applied is that air masses are always in hydrostatic equilibrium and there is no vertical acceleration of air parcels. This assumption may no longer be valid when the resolution is refined.

An important caveat to be kept in mind is that climate models, global or regional, can rarely reproduce the observed mean climate with the desired accuracy. All climate models are burdened by systematic errors: the mean temperature or mean precipitation is systematically lower or higher than the corresponding observed measured data, for example. This is denoted as bias. For instance, Fig. 4.2 shows the

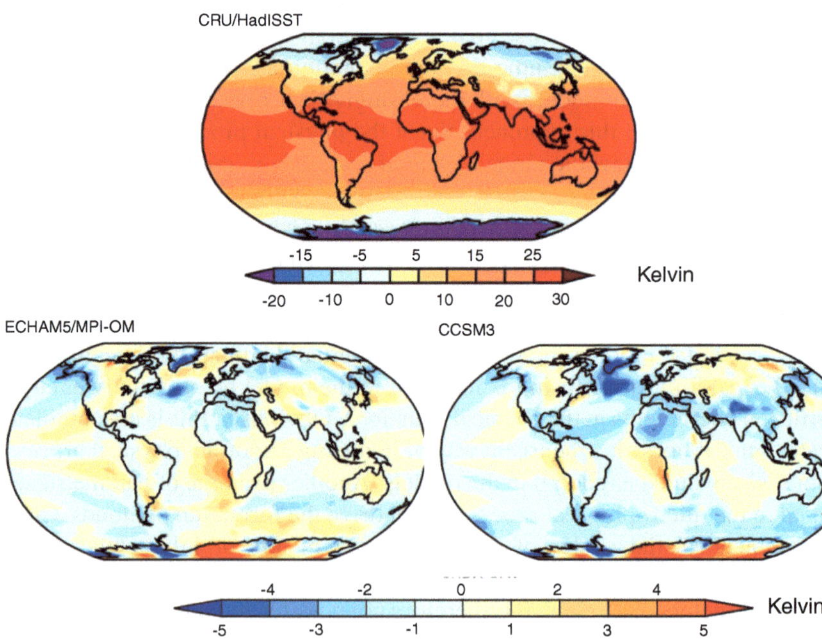

Fig. 4.2 *Top*: Mean annual near-surface temperature for the present climate. *Bottom*: Differences between the simulated and the observed temperatures in two global climate models, denoted as model temperature bias

observed global annual mean near- surface temperature, and the mean differences between this observed pattern and the simulated pattern by the models ECHAM5-OM (Max-Planck Institute for Meteorology) and the model CCSM3 (National Centre for Atmospheric Research). Although the differences between simulations and observations are just within a few degrees Kelvin, these differences result in a large bias, since the simulated climate change by the end of the twenty-first century could also be in the range of a few Kelvin. However, this systematic error does not necessarily mean that the simulation of *climate change* under perturbed external forcings is incorrect. It may well occur that the increase in temperature caused by changes in one of the external drivers is not very dependent on the *mean* temperatures. This will be more realistic if the temperature changes are small. However, it is of course desirable that climate models simulate the observed climate as closely as possible, and a lot of efforts are invested to try and achieve this. The fact that this has not been satisfactorily achieved is an indication of the daunting difficulty of this task.

4.3 Climate Simulations for the Past and Future

Climate simulations for future periods are obviously of interest for estimating possible climate changes due to increasing concentrations of greenhouse gases in the atmosphere. But there are also good reasons to simulate the climate of the past. Simulations of the past climate can be helpful in interpreting the evidence obtained from proxy data, such as tree rings, analyses of ice cores or lake sediments, among others (see Jones et al. 2009 for a comprehensive review of proxy-based reconstructions in the past centuries). The proxy evidence is usually fragmentary, describes local conditions and the proxy indicators are influenced not only by environmental conditions but also by other non-climatic factors. For instance, tree growth is also modulated by competition among trees, infections, fire, and other factors, so that climate information is always contaminated by non-climatic factors. Climate simulations can provide a global and dynamically consistent picture of past climates and can therefore help to understand and put in a broader context the evidence provided by different types of proxies and proxies from different locations. For instance, if a set of proxies indicates warm conditions in one region and other sets indicate wetter conditions in other regions for the same time period, climate simulations can provide clues about whether both types of climate conditions may be caused by large-scale atmospheric circulation patterns and are reasonable in terms of physical climatology.

The second important application of simulating past climates is the validation of climate models themselves. Climate changes occur at slow paces, not at the daily or weekly time scales on which weather variations occur, so it is very difficult to ascertain whether or not a particular climate model simulates climate changes realistically. Using the present for validating climate models would mean waiting 20 or 30 years until the climate change response is well established in the observations, and then compare the observations with the simulations. The rationale of paleoclimate simulations is to simulate past recent periods and perform the same type of comparison. It turns out that the twentieth century, which is relatively well covered by observational data sets, is not long enough to discriminate between good and less good climate models. Longer periods, in particular with clearly colder or warmer temperatures than at present, are best for model testing. This is because the differences in mean climates are not blurred by the ubiquitous internal variations in periods of strong climate anomalies, as discussed previously. On the other hand, these periods which form the base for comparison, should ideally be covered by climate information from proxy data as well. Most of the simulations of past climates so far have been performed for the Last Glacial Maximum, about 21,000 years ago, for the Holocene Optimum about 6,000 years ago, and for the past millennium.

To simulate the past climate, it is necessary to estimate how the external climate drivers have changed in the past. For the three periods mentioned above, three types of climate drivers are important: the variations in the orbital parameters of the Earth (perihelion, obliquity and eccentricity: changes in all three are important for the last Glacial Maximum, changes in the perihelion are important only for the Holocene Optimum and the past millennium). The perihelion, which denotes the closest position of the Earth to the sun in a calendar year, varies with a period of roughly 18,000 years. Currently, it occurs at the beginning of January, but about 10,000 years ago, it occurred in July. Consequently, summers in the Northern Hemisphere 10,000 years ago tended to be warmer than today, and winters cooler. In fact, all available proxy information available indicates that summers have become cooler at centennial time scales, throughout the Holocene until present. An interesting test for a climate model is to check whether this cooling trend in Northern Hemisphere temperatures can be well reproduced by a climate model driven by variations of the position of the perihelion.

For the past millennium, variations in the perihelion play a small role, but the influence of other forcings became more important. Solar irradiance – the total energy irradiated by the sun – varies at decadal and multi-decadal time scales. These variations can be reconstructed from the concentrations of cosmogenic isotopes (^{10}Be in ice cores or ^{14}C in tree-rings), since the production rate of these two isotopes in the stratosphere is modulated by the strength of the solar open magnetic field. Another important external forcing are the concentrations of greenhouse trace gases in the atmosphere, CO_2 and CH_4. Past concentrations of greenhouse gases can be estimated by analyzing air bubbles trapped in polar ice-cores when deposited snow was slowly compacted under its own weight to form ice.

Other forcings with a more regional character are changes in land use by human action. As discussed before, the clearance of forests to grow crops changes the energy, momentum and water fluxes between the land surface and the atmosphere. Forests tend to have less reflectivity than farmland and therefore absorb more of the incoming sunlight, offer more resistance to wind flow, and if water is available, trees may transport a lot of soil water to the atmosphere via evapotranspiration.

The construction of simulations for the future climate requires knowledge of the trajectory of future external forcings, which obviously is very uncertain. Thus, scenarios of future forcings need to be prescribed in model simulations. These simulations, although intended to represent future climate, cannot be considered as predictions *per se*, as they depend on assumptions leading to the forcing scenarios. The climate simulations for the future are based on various so-called emission scenarios (to quantify the future climate forcing arising from greenhouse gases). Emissions scenarios are constructed using complex considerations of social and economic development to estimate future levels of anthropogenic carbon emissions. The scenarios are condensed in the Special Report on Emissions Scenarios (SRES, IPCC 2000). In total, they comprise 40 scenarios for potential future anthropogenic emissions of greenhouse gases, grouped into scenario families according to different assumptions on future environmental policies and the level of globalization. An important point is that no relative probability is attached to any particular scenario, i.e. all are considered equally probable and they are not intended to cover all possible future emissions paths. Therefore, the ensembles of simulations carried out using these emissions scenarios with different climate models are not probabilistic predictions either. They indicate a range of future climate evolutions conditioned on the present knowledge of what we may consider plausible paths of future emissions.

Keeping these caveats in mind, Fig. 4.3 displays the simulated global and annual mean temperatures simulated by the model ECHO-G over the past millennium and in the twenty-first century, driven by reconstructions of the past forcings and a scenario of future external climate forcings. The most prominent feature is the increase in temperature in the twenty-first century compared to the variations in the past 1,000 years. This increase in temperature is associated with the rapid increase in concentrations of anthropogenic greenhouse gases under the SRES scenario (A2). This scenario represents a very rapid growth in emissions over the twenty-first century. The overall picture remains qualitatively unchanged when other estimations of past external forcings are used. The future temperature increase would be scaled down accordingly if more optimistic future emissions scenarios were assumed. Figure 4.3 also displays the simulated evolution of the annual temperature and precipitation in the Baltic Sea area over the same period. These curves show a very similar behaviour compared to the global mean annual temperature. At the beginning of the millennium, temperatures tended to be warmer than later in the middle of the millennium. The exact temperature level in this early period (the Medieval Warm Period), relative to the present temperatures, is difficult to establish because it is affected by the initial conditions in the model simulation, which are not really known. The middle centuries in the millennium are characterized by lower temperatures than today, reaching a minimum at roughly AD 1700. These central centuries are denoted in the literature as the Little Ice Age. Subsequently, global and Baltic Sea regional temperatures tend to rise up to the present level, and later in the twenty-first century, they clearly attain values outside the natural variations over the last millennium. The evolution of the Baltic Sea regional temperature is qualitatively similar to that of the global temperature, but

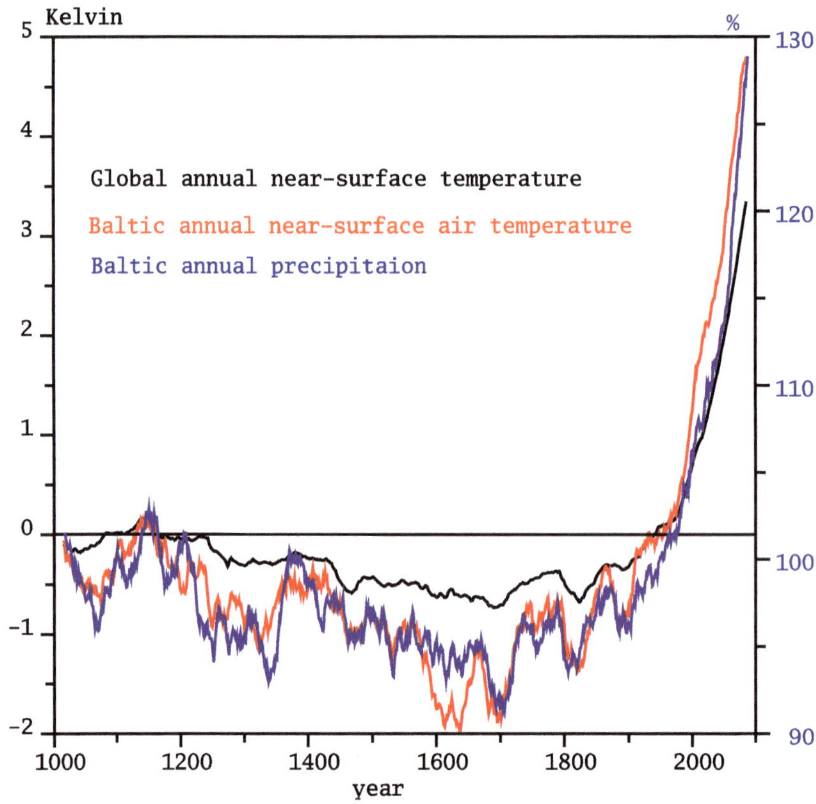

Fig. 4.3 Global annual near-surface air temperature, annual Baltic Sea regional temperature and annual Baltic Sea regional precipitation, simulated by the global climate model ECHO-G in the past millennium and in the twenty-first century according to scenario A2. The depicted temperatures are deviations from the simulated 1900–1990 mean, and precipitation is the ratio to the mean precipitation in 1900–1990. All series have been smoothed with a 31-year running mean filter

the amplitude of variations is larger at the regional scale. Usually, high latitude regions display larger temperature variability and also larger temperature responses to external forcings. This phenomenon is called polar amplification and is partially explained by the ice-albedo feedback, i.e. reduction in ice and snow cover accentuates warming due to the increased absorption of solar radiation. Annual precipitation in the Baltic Sea area is closely associated with the evolution of temperature. The most plausible explanation is that precipitation is mostly controlled by the specific humidity at these latitudes, as this area is always crossed by the Atlantic mid-latitude cyclones, so that a possible shift of the storm tracks due to climate change does not strongly affect the number of cyclones flowing over the Baltic Sea area. This is confirmed by a previous analysis of extratropical storm activity in the simulations (Fischer-Bruns et al. 2005).

This simulated behaviour is due to the evolution of the external forcings. In the early centuries, solar irradiance was stronger than during the Little Ice Age, during which also the frequency of volcanic eruptions increased. Later, solar irradiance increased from AD 1700 onwards and, in addition, the concentration of greenhouse trace gases rose due to anthropogenic activities. In the twenty-first century, this latter contribution is the only one that contributes to the increase in external forcings, since solar irradiance was kept constant at the level of 1990 (it is currently not possible to predict the evolution of the solar irradiance beyond a few years). Furthermore, no volcanic eruptions are included in the simulations, as these are also unpredictable. It is, however, quite obvious, that if the variations of the natural forcings remain within the ranges of the past millennium, their contribution would be negligible in comparison with the contribution from anthropogenic greenhouse gas forcings, under these scenarios of future emissions.

How reliable are these simulations over the past millennium? This is a difficult question to answer, since climate reconstructions based on climate proxies are also burdened with considerable uncertainties, which will be not explained here in detail. The interested reader may consult the review paper by Jones et al. (2009). The available reconstructions of global or northern hemisphere temperatures differ due to their use of different proxy series, but most importantly due to the use of different statistical *calibration* methods. The proxy indicators, such as tree rings, have to be translated in terms of seasonal temperature or precipitation, and this is usually accomplished by comparing these indicators with observations in an overlapping period in the twentieth century. Linear regression methods or more sophisticated methods are then used to estimate transfer functions that translate the variations of tree ring width in terms of local environmental variables. However, the connection between both is sometimes not very close, so that other factors must also be affecting the variations of the proxy records. For instance, correlations between instrumental records and proxy records in the twentieth century are in the order of 0.5, and only very good proxy records attain higher correlations. These loose connections between proxy records and the local climate causes uncertainties in the "pre"diction of temperature back in time when the full proxy time series is used. Furthermore, the proxies are representative of a relatively small area or season, and this has to be extrapolated to estimate a global or northern hemisphere mean, based on a limited number of proxies. Thus, the assessment of accuracy in the paleo model simulations is carried out using data sets which are, themselves, inaccurate to some extent.

The possible impacts of future climate change are most usefully estimated by climate models on the regional scales. As mentioned before, the skills of global climate models are reduced at high spatial resolutions. For instance, let us assume for the moment that a particular climate model is not able to represent the climate of Northern and Southern Europe properly. If these deviations are independent, they will tend to cancel themselves out when considering a continental or large-scale average. This large-scale average would then conceal a bad performance at regional scales. It is therefore interesting to examine the capabilities of climate models to simulate changes at regional spatial scales. One problem that has to be taken into account is that, for the same reason that non-systematic deficiencies tend to average out in large-scale means, the climate variations at regional scales are more strongly influenced by internal stochastic variations than global or large-scale averages.

A clear example is provided by winter temperature in Northern Europe. It is well known that this variable is strongly modulated by the intensity of the westerly winds over the North Atlantic - stronger winds than normal bring milder winters, whereas weaker westerly winds or even easterly winds are associated with colder winters. This happens independently of variations in the external forcings and therefore, is a result of internal stochastic variation. The evolution through time of the westerly winds in real climate and in a model simulation will be different in principle, although their statistical properties – mean and standard deviation – would ideally be the same. The consequence is that the trajectories of the winter temperatures in Northern Europe over the twentieth century, or some other comparable period of time, will considerably disagree in the observations and in the simulation, even if the climate model is perfect. Only at time scales in which the external forcing exerts a strong influence on climate, observations and simulations will potentially be comparable. The time scale will be dictated by the need to average out the influence of internal stochastic variations.

Despite this caveat, a comparison between paleoclimate reconstructions of the temperature in Northern Europe and the results of a climate simulation can be instructive. Such simulations can indicate at which time scales we can expect agreement between both, and whether or not an agreement exists on longer time scales. One comparison is provided by Fig. 4.4. This figure shows a reconstruction of winter-spring temperature in Stockholm over the past five centuries, together with the results of three climate simulations: two performed with the global climate model ECHO-G and one performed with a regional climate model RCAO, driven by one of the ECHO-G simulations (Zorita et al. 2010). The temperature reconstructions are based on information from historical administrative documents from the port of Stockholm, which registered the dates of the first sailings in spring every year. This date is strongly influenced by the conditions favourable for

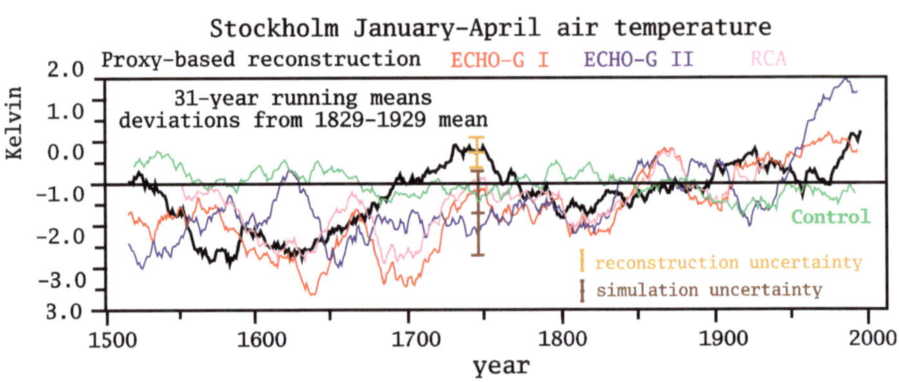

Fig. 4.4 Reconstructions of the past winter-to-early spring air temperature in Stockholm in the past 500 years, compared to the output of two climate simulations using the model ECHO-G, and one climate simulation using the regional model RCA. In addition, the temperature obtained in a control simulation is also shown. The uncertainty ranges in the reconstructions and in the simulations are also indicated (From Zorita et al. 2010)

navigation in the port, since usually the port was frozen over the winter and not navigable. By applying statistical methods for calibration analogous to other proxy series, it was possible to derive the mean temperature in winter-early spring in Stockholm. The two global simulations – results from one of which are seen in Fig. 4.3 – were performed in identical ways – using the same model and the same external forcing – the only difference being the initial conditions in the year 1,000. Both simulations, therefore, provide a rough indication of the expected spread in temperatures due to internal stochastic variations in the regional climate. Figure 4.4 shows that both simulations do not agree at decadal and multi-decadal time scales, and only the long-term warming trend across past centuries is consistent between the two. A control simulation (also included in the figure), in which the external forcing was kept constant, does not show a comparable trend over the past centuries. This is a strong indication that the temperature trend simulated in the two forced simulations is due to the external forcing. The trend also agrees quite well with the trend derived from the temperature reconstructions based on the documented data, supporting the skill of the models in simulating the long-term response of the Stockholm temperature. Note that reconstructions and simulations were accomplished by completely independent groups of researchers and that no information was exchanged between them. Under these conditions, the agreement is quite remarkable.

However, the change in summer temperatures from the past centuries into the present in the Baltic Sea area is overestimated in this model, compared to proxy-based reconstructions (Hansson and Omstedt 2008), since this model does not include the effect of tropospheric aerosols, a climate forcing factor that presumably has contributed to dampen the effect of greenhouse gas warming in the second half of the twentieth century.

On the other hand, disagreements exist concerning the magnitude of variations at shorter decadal time scales. For instance, the decades around AD 1730 were relatively warm according to the temperature reconstruction, with large and persistent deviations from the long-term centennial trend. This type of excursion does neither appear in the forced simulations, nor in the control simulation. Furthermore, the external solar forcing is not consistent with these warm decades. This may be an indication that the climate model is not able to generate large internal variations at the regional time scale, a deficiency that should be further investigated.

This example illustrates nicely how the joint analysis of model simulations and proxy-based reconstructions can help improve climate models and indirectly reduce the uncertainties in the climate projections for the future. These uncertainties are further discussed in the next section.

4.4 Uncertainties in Simulating the Future

A simulation of the future climate can be considered as a tool to translate the scenarios of possible future concentrations of greenhouse gases into changes in temperature, precipitation and other climatic variables. Therefore, they are not *per se* climate predictions, but rather answers to the questions posed by the scenarios themselves – a *what if* question.

For the objective of estimating future climate scenarios on global scales and on the scale of the Baltic Sea area, several sources of uncertainties have to be considered. The major source of uncertainty is, obviously, the future path of concentrations of greenhouse gases. This path is highly uncertain, although some guidance is provided by the integrated assessment models that incorporate representations of the global socio-economic systems and its interaction with climate itself. Different policies, demographics and technological developments may give rise to quite different future concentration paths. Furthermore, the eventual mitigation response of human societies to a changing climate also modifies climate change itself.

The 40 SRES scenarios are grouped into four families A1, A2, B1 and B2. The 'two' families assumed that the future world will not develop into 'integrated' geopolitical structures and thus it will remain 'more regional'. Within the 'one' family the world will converge in supraregional, more integrated, geopolitical structures and the level of coordination will be higher. The 'B' family assumes a more rapid implementation of more ecologically friendly policies, whereas in the 'A' family this will not occur as fast. Within each family, further groups have been envisaged according to more detailed technological assumptions about energy generation and demographics. Thus, the A2 scenario is one of the most pessimistic, although some of its underlying assumptions about demographic development (15 billion people by AD 2100) will not come to pass according to more recent UN population projections (a median scenario of nine billion by AD 2050, decreasing thereafter). The B1 scenario (an integrated world with a rapid implementation of new and ecological technologies) is more optimistic, and the scenario A1B (an ecological world based on regional solutions and balanced energy production) lies roughly in the middle of all SRES scenarios.

Other, more physically based sources of uncertainty are climate models. Although all climate models are based on the same set of physical and chemical laws, they do not provide the same response to the same prescribed path of future concentrations of greenhouse gases. The spread of the modelled response is not negligible. Even for highly simplified theoretical evolutions of greenhouse gas concentrations, for instance a sudden doubling of the concentration of CO_2 allowing the climate system to reach an equilibrium after several centuries, different models indicate different increases in the global mean annual temperature. The magnitude of this change is denoted as the *climate sensitivity* and ranges between 2.1° and 4.4° for the climate models included in the last IPCC Fourth Assessment report (IPCC 2007).

There may be different reasons for these differences, but some are more likely than others. The main reason lies in the internal reaction to the prescribed external forcing, the feedbacks. When the external forcing is increased and the temperature tends to rise accordingly, several other changes may accompany this temperature rise. Evaporation from the oceans will tend to increase, and since water vapour is a very potent greenhouse gas itself, this will tend to reinforce the initial temperature rise. The characteristics of cloud cover will also change. Clouds have a very large potential to modulate the energy fluxes in the atmosphere and at the surface of the Earth. All types of clouds reflect visible short-wave solar radiation and absorb invisible infra-red radiation, and therefore may have a moderating or reinforcing effect of the initial temperature perturbation. Low stratiform-cumulus clouds in the tropics mostly reflect solar radiation and their greenhouse effect is small, whereas cold high cirrus clouds that mostly contain ice crystals are almost transparent to visible solar radiation but are very potent absorbers of infra-red radiation. Their greenhouse effect is thus very strong. Depending on how cloud cover and cloud height changes with temperature in each model, the final temperature response, including the initial perturbation and the reinforcements or moderating factors, is different in each model. Clouds are the most important single factor explaining the different temperature responses across models.

An illustration of the uncertainties in the simulation of temperature change under the scenario A1B is provided in Fig. 4.5. This figure displays the mean

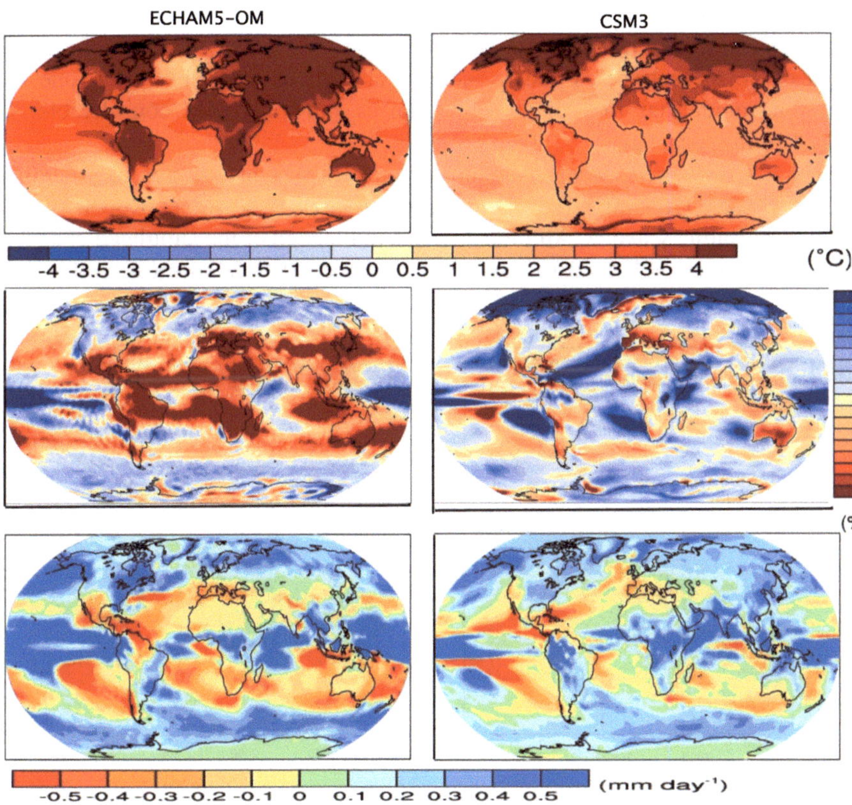

Fig. 4.5 Changes in annual temperature (Kelvin, *top*), annual total cloud cover (%, *middle*) and annual precipitation (mm per day, *bottom*), simulated by the models ECHAM5-OM and CCSM3 used in the Fourth Assessment Report of the IPCC, and driven by scenario A1B of future anthropogenic emissions of greenhouse gases. The maps display the differences between the 2080 and 2099 mean and the 1980–1999 mean

annual temperature difference between the end of the twenty-first century and present, as simulated by two climate models: ECHAM5-OM from the Max-Planck Institute for Meteorology in Hamburg, and the model CCSM3 from the National Centre for Atmospheric Research in Boulder. The figure additionally shows the simulated changes in total cloud cover and annual precipitation as simulated by both models.

In general, the model CCSM3 simulates a smaller temperature change than the model ECHAM5-OM. The model climate sensitivities are 2.7 and 3.4 K, respectively. When we consider the simulated changes in total cloud cover – i.e. including all types of clouds – the differences between both simulations are evident. The model ECHAM5-OM tends to simulate stronger reductions in cloud cover than CCSM3. In large tropical areas, the simulated changes by both models have opposite signs. This is also reflected in the simulated change in precipitation. The precipitation reductions simulated by ECHAM5-OM are more severe, and some areas also have opposite signs to those simulated by CCSM3, as in the Monsoon region and Southern Africa. The large-scale spatial patterns of temperature and precipitation changes are, however, reasonably consistent. Temperatures tend to increase stronger at high-latitudes and over the continents. Precipitation tends to decrease in the subtropics and to increase at high latitudes. However, one has to bear in mind, that the regional differences may be quite considerable, as indicated in the next figure.

Figure 4.6 shows the paths of annual temperature and winter precipitation changes simulated by all models in the last Fourth Assessment Report of the IPCC for the Baltic Sea area. All models agree in the sign of the simulated changes: annual temperature will tend to rise relative to the 2000–2010 mean and winter

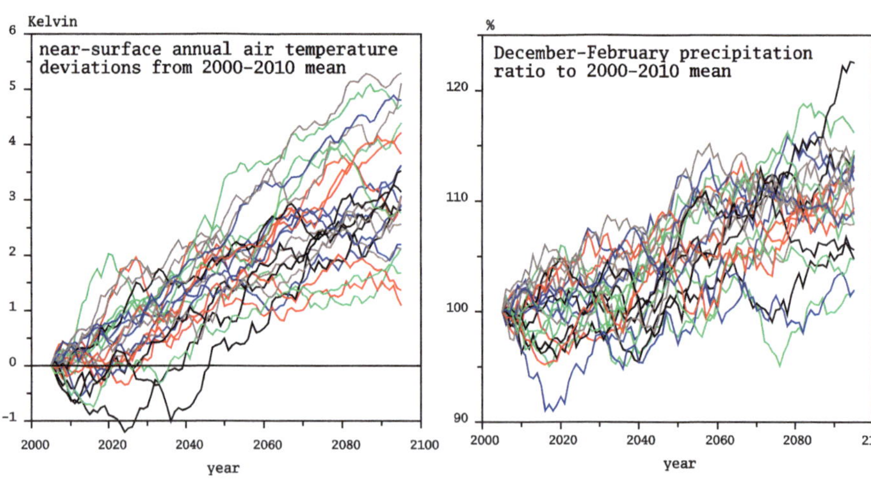

Fig. 4.6 Changes in annual temperature and December–February precipitation in the Baltic Sea region simulated by 24 models used in the fourth Assessment Report of the IPCC, and driven by scenario A1B of future anthropogenic emissions of greenhouse gases. The series were smoothed by a 11-year running mean filter

precipitation will tend to increase relative to the mean of the same period. However, the spread in the simulations is quite large. In the case of temperature, the range of change is about 1.2–5 K, and in the case of precipitation the range is about 0% and 20% increase. The bulk of the simulations appear to cluster around 3 K of temperature change and 10% increase of winter precipitation, but as mentioned before, we cannot attach probabilities to this range unless we assume that all models are equally good and that the 24 model suite completely covers the range of all possible models.

4.5 Specific Problems for the Baltic Sea Area

Global climate models are a representation of our knowledge on the physics of the atmosphere and the ocean. Both subsystems of the global climate system interact with each other, exchanging energy, momentum and matter, and modern climate models also represent this coupling. The limited resolution of global climate models, however, is a problem when we are interested in the Baltic Sea area. Although in principle, it should be possible to *downscale* the output of global models with the help of regional climate models, as presented in the previous section, regional models usually only represent the atmosphere. They are driven at their lateral boundaries by the output of global models and by the ocean surface conditions also simulated by global models (or taken from observations if the period of simulations is the present). The additional problem in the Baltic Sea, and indeed in all areas with complex coastlines, is that the surface boundary conditions simulated by global models are not realistic. For instance, if the connection between the Baltic Sea and the North Sea in a global model is about 400 km, the heat transport by ocean currents will be unrealistically large, and the Baltic Sea surface temperatures provided by the global model will not be correct. This deficiency can only be solved by the use of a *coupled atmosphere-ocean regional model*. Such a model will then be driven only by lateral conditions in the atmosphere and in the ocean simulated by the global model, but it would simulate its own surface temperature over water, ice, and other surface fluxes within the Baltic Sea area. An additional advantage in case of the Baltic Sea basin is that in these coupled regional models, run-off is much better simulated. Run-off is an important climate variable, to a large extent determining the salinity of the Baltic Sea.

Unfortunately, a regional coupled model is a quite complex piece of simulation software, and for the Baltic Sea area not many models of this kind exist. The best known and more widely used is the RCAO model of the Swedish Meteorological and Oceanography Institute (SHMI). The Max-Planck Institute for Meteorology has also developed a coupled model for the Baltic Sea, Baltimos. This model has been used to simulate the future climate of the Baltic Sea under different emission scenarios. An additional source of uncertainty in this case is caused by the global climate model used to drive RCAO. Different global models would, in general, provide different lateral boundary conditions, as was demonstrated in Fig. 4.6. To estimate the full range of uncertainties for a particular area as the Baltic Sea, the analysis should comprise a battery of simulations with different global models, driving different regional models, and themselves being driven by different emissions scenarios. The computing effort to accomplish this estimation of uncertainties is huge and it has been first tackled in the framework of large projects funded by the European Commission, like Ensembles. Here, only some examples of this kind of simulations are presented.

Figure 4.7 shows the simulated changes in surface temperature and vertically integrated salinity in the last three decades of the twenty-first century, using the emissions scenario A2. Although the driving global model does not belong to the IPCC suite in the Fourth Assessment report, and therefore it is not included in Fig. 4.6, the main characteristic of the global simulation is similar to the results obtained with the model ECHO-G and ECHAM5-OM. All three models indicate an increase in annual precipitation in this area. On the other hand, the increase in temperatures simulated by the end of the twenty-first century should be associated with increased evaporation. This leaves the question, which factor will be dominating in the simulated changes of Baltic Sea salinity? This is one example of a complex and important question in which a regional climate model such as RCAO becomes very valuable. The RCAO model indicates that the increase in precipitation will dominate under this scenario and integrated salinity will tend to decrease. Furthermore, the oceanic sub-model of the RCAO coupled model also provides an estimation of the spatial pattern of temperature and salinity changes at very small spatial scales. This can also be used, for instance, to estimate ice cover changes and its spatial distribution. In addition, the output of the RCAO can be fed into ecological models of the Baltic Sea primary production and food web, thus

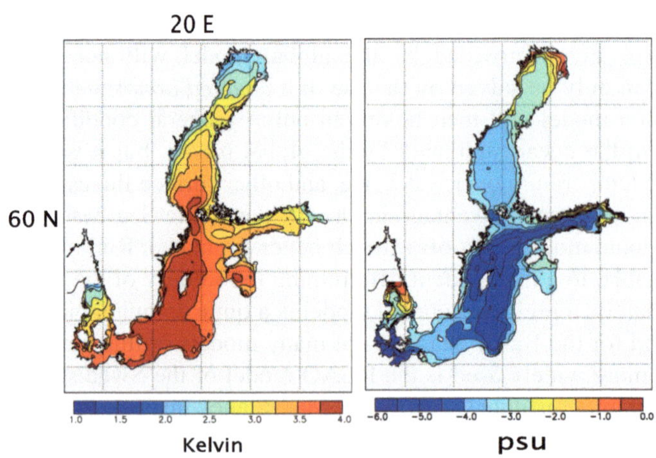

Fig. 4.7 Changes in the Baltic Sea in the period 2070–2100 as compared to present conditions, as simulated by the regional coupled atmosphere-ocean model RCAO, and driven by the global model ECHAM4-OPYC3 under emissions scenario A2. *Left*: Sea-surface temperature; *Right*: Salinity, integrated through the vertical water column (From Meier 2006)

bridging the gap between the large-scale climate changes and climate impacts at a very small scale. Other models will benefit from the progress in regional climate modelling achieved in the recent years.

The uncertainties in the simulations performed by regional climate models should not be overlooked. Some important processes that may influence future run-off changes, and therefore future Baltic Sea salinity, are not perfectly included in the model. Vegetation cover may change in the future due to climate change itself, which will in turn influence evapotranspiration and run-off. Evapotranspiration may increase due to warmer temperatures and better growing conditions for plants at these latitudes in the future. On the other hand, higher CO_2 concentrations in the atmosphere may lead to an increased water efficiency, since plants would be allowed to keep their stomata (tiny leaf pores allowing gas exchange) open for shorter periods. Other studies which have also analysed the relationship between temperature, run-off and salinity in the Baltic Sea in the past centuries (Hansson and Omstedt 2008) have reached the conclusion that, if this relationship remains unchanged in the future, salinity in the Baltic Sea as a whole will tend to increase. These results illustrate that caution is always needed when interpreting future projections.

Acknowledgments I am thankful to Anders Omstedt and one anonymous reviewer, as well as Mary Gagen and Birgit Hünicke, for their very constructive comments and their careful correction of previous manuscript drafts. I profusely used the Fourth Assessment Report of the Intergovernmental Panel on Climate Change (IPCC) and its data bank storing climate simulations at the Lawrence Livermore National Laboratory.

Appendix

A Brief Description of the Climate Models Mentioned in this Chapter

ECHO-G is a *global* coupled atmosphere-ocean model. It is composed of the atmospheric model ECHAM4 and the ocean model HOPE. ECHAM4 is a spectral model with a horizontal resolution of about 3.75° × 3.75° and 19 levels in the vertical direction. HOPE is a finite-difference model with a horizontal resolution of about 2.8° × 2.8° and 20 levels in the vertical direction. Due to the relatively coarse horizontal resolution, the fluxes of energy and momentum between both models contain errors that would cause a small but persistent long-term drift of the coupled model. This is quite detrimental for long simulations and to avoid this drift, a flux adjustment is applied to couple both models.

ECHAM4-OPYC is also a *global* coupled climate model as ECHO-G but with another ocean sub-model. In the place of HOPE, another ocean model OPYC is coupled to ECHAM4. A special characteristic of OPYC is that its levels in the vertical are defined at surfaces of constant density, and not of constant depth.

ECHAM5-OM is a *global* climate model that consist of the atmospheric model ECHAM5 (a newer version of ECHAM4) and the ocean model OM. The horizontal resolution of the model ECHAM5 is typically $1.9° \times 1.9°$. It does not require flux adjustment as ECHO-G does. In the set-up used for the IPCC simulations, this model includes a vegetation model and a carbon cycle model.

CCSM3 is a *global* atmosphere-ocean climate model developed at the National Centre for Atmospheric Research in Boulder (USA). The atmospheric sub-model is a spectral model with a resolution, for the IPCC simulations, of about $2.8° \times 2.8°$, and 20 levels in the vertical direction. It does not require a flux adjustment.

MM5 is a *regional* atmospheric climate model. It needs to be nested in another global model, i.e. it requires values of the temperature, humidity and wind every 6 h at the boundaries of the simulation domain by a global climate model, or by meteorological re-analysis. Its resolution may be chosen by the user within a certain permissible range. Typically, resolutions of about 50×50 km are used, but simulations with finer resolutions, for instance half these values, are also performed. An interesting property of this model is that it allows for the so called double nesting set-up. In this set-up, a smaller model domain with a higher resolution is nested within a larger domain with a somewhat coarser resolution, which in turn is nested in a global climate model. In the example presented in this chapter for simulations of the climate of the Iberian Peninsula, the high-resolution inner domain covered the Iberian Peninsula with a resolution of 45×45 km and this domain was embedded in the outer model domain, covering the North Atlantic-West European sector with a resolution of 120×120 km. This model allows to choose among different parametrizations of sub-grid processes involved in cloud formation, convection, etc., so that different results may be obtained depending on the values of these parameters. In general, these parameters are selected according to the climatic region that is simulated.

WRF is essentially a newer version of the *regional* model MM5.

RCAO is a *regional coupled* atmosphere-ocean model developed at the Swedish Rossby Centre. It comprises a regional atmospheric model RCA and an ocean regional model RCO that includes a sea-ice model. It has been mainly applied for the Baltic Sea basin. The atmosphere resolution is similar to those in other regional atmospheric models such as MM5. The ocean resolution may typically be of the order of 6 nautical miles, with 40 levels in the vertical direction.

BALTIMOS is, like RCAO, a *regional coupled* atmosphere-ocean model, based on the regional climate model REMO. It has been developed at the Max-Planck-Institute for Meteorology and one of its main applications has been to simulate North Sea water intrusions into the Baltic Sea.

References

Fischer-Bruns I, von Storch H, Gonzalez-Rouco J, Zorita E (2005) Modelling the variability of midlatitude storm activity on decadal to century time scales. Clim Dyn 25:461–476

Hansson D, Omstedt A (2008) Modelling the Baltic Sea ocean climate on centennial time scale; temperature and sea-ice. Clim Dyn 30:763–778

IPCC (2000) Special report on emissions scenarios. www.grida.no/climate/ipcc/emission/index.htm

IPCC (2007) Contribution of Working Group 1 to the Fourth Assessment Report. http://www.ipcc.ch/publications_and_data/ar4/wg1/en/contents.html

Jones PD et al (2009) High-resolution paleoclimatology of the last millennium: a review of current status and future prospects. Holocene 19:3–49

Meier MHE (2006) Baltic Sea climate in the late twenty-first century: a dynamical downscaling approach using two global models and two emission scenarios. Clim Dyn 27:39–68

Rummukainen M (2010) State-of-the-art with regional climate models. Wiley Interdiscip Rev Clim Change 1:82–92

Wilby RL, Charles SP, Zorita E, Timbal B, Whetton P, Mearns LO (2004) Guidelines for use of climate scenarios developed from statistical downscaling methods. Supporting material of the IPCC Third Assessment Report. http://www.ipcc-data.org/guidelines/dgm_no2_v1_09_2004.pdf

Zorita E et al (2010) European temperature records of the past five centuries based on documentary/instrumental information compared to climate simulations. Clim Chang. doi:10.1007/s10584-010-9824-7

Chapter 5
Impacts of Climate Change, Including Acidification, on Marine Ecosystems and Fisheries

Keith Brander and Jon Havenhand

5.1 Introduction

Marine ecosystems have always been affected by changes in climate at timescales from decades to millions of years. Since the industrial revolution in the nineteenth century the increase in greenhouse gases (GHG) has caused an accelerating rise in global temperature whose effects on marine biota can be detected at individual, population and ecosystem level. The rising level of CO_2 and consequent acidification of the oceans is having an impact on metabolism and calcification in many organisms, with damage to vulnerable ecosystems, such as coral reefs, already occurring. The pH of the oceans is already lower now than it has been for the past 600,000 years.

We depend on the oceans and coastal seas for many ecosystem services (see Table 5.1) and some of these services are being damaged and degraded by climate change. Can we identify areas and ecosystems that are particularly vulnerable? How quickly will climate affect marine ecosystem services? How great is the impact likely to be under different scenarios of climate change? Do we have monitoring systems in place that will warn us of impending changes and provide information that can be used to respond? What are the options for adapting to climate change and for mitigating GHG emissions in shipping, fisheries and other maritime activities? Credible responses to these questions require improved understanding of processes. Experiments and field work on all levels of biological organisation will provide this.

K. Brander (✉)
Technical University of Denmark, Charlottenlund Slot, Jægersborg Allé 1, Charlottenlund 2920, Denmark
e-mail: kbr@aqua.dtu.dk

J. Havenhand
Department of Biological & Environmental Sciences, University of Gothenburg, 452 96 Tjärnö, Strömstad, Sweden
e-mail: jon.havenhand@bioenv.gu.se

M. Reckermann et al. (eds.), *Climate Impacts on the Baltic Sea: From Science to Policy*, Springer Earth System Sciences, DOI 10.1007/978-3-642-25728-5_5,

Table 5.1 Classification of ecosystem services (Based on the Millennium Ecosystem Assessment 2005)

Ecosystem service type	Ecosystem service
Provisioning	Food
	Fibre
	Medicine
	Cosmetics
Regulating	Carbon sequestration
	Water regulation
	Climate regulation
	Coastal protection
	Water purification
	Disease and pest control
Cultural	Spiritual values
	Aesthetic value
	Intrinsic value
Supporting (the other service types)	Primary and secondary production
	Biodiversity

Climate change is not the only pressure that humans impose on the marine environment and in the short term it is probably not the one with the greatest impact on biodiversity and productivity. Fishing, habitat degradation, pollution and species introductions also have many undesirable consequences and interact with each other. In coastal and inland waters there are additional effects due to changes in land use, coastal defence, damming, flood control and alteration of waterways.

The Baltic Sea is a particularly well studied enclosed sea, with a range of ecological conditions from freshwater to marine. The ecosystems are sensitive to certain aspects of climate, such as change in salinity, and some of the lessons concerning the interaction between natural and human-induced pressures can provide useful insight for the management of this and other marine areas.

Climate change can act directly on productivity of commercial fish species by altering growth, reproduction and other aspects of life history either positively or negatively. It can also act indirectly by enhancing or suppressing prey, predators and pests, with consequences for the structure and resilience of the whole ecosystem. The processes and pathways by which biological production is transferred through marine food-chains are longer and more complex than in terrestrial agriculture and the scope for experiments and research is far less. Fisheries production depends on net primary production (NPP) and the transfer of this production through the food chain to human consumption. Future NPP may increase in some high latitude regions due to warming, decreased ice cover and changing pressure and wind-fields. The dynamics in low latitude regions are governed by different processes and NPP may decline due to reduced vertical mixing of the water column and hence reduced recycling of nutrients. The effects of decreasing ocean pH on NPP are as yet unknown. Recent changes in distribution and productivity of a number of fish species can be ascribed with high confidence to regional climate variability, such as the El Nino Southern Oscillation (ENSO). The effects of fishing and of climate interact, because fishing reduces the age, size and geographic diversity of populations and the biodiversity of marine ecosystems, making both more sensitive to additional stresses, such as climate change.

The frequency and intensity of extreme climate events, such as floods, droughts, heat waves and storms, is likely to have a major impact on future fisheries production in both inland and marine systems. Reducing fishing mortality in the majority of fisheries, which are currently fully exploited or overexploited, is the principal feasible means of reducing the impacts of climate change.

5.2 Global and Regional Climate Change

Variables for ocean climate are of course different from those for the atmosphere, but the atmospheric variables (temperature, precipitation, winds) also affect ocean processes. A listing and classification of the main ocean climate variables is given in Table 5.2.

The rate of increase in global air temperature has accelerated over the past century (Fig. 5.1) and is expected to continue to do so. The recent rate of increase in sea surface temperature (SST) in the NE Atlantic (1980–2005) has been quite varied, but the North Sea and Baltic Sea have warmed much faster than the global rate, with a rise in temperature of 1–1.5°C over the 25 year period (Fig. 5.2). The fact that warming in this region has been so much faster than the global average raises two important issues. First, impacts on marine ecosystems are likely to be greater and more easy to detect in this region than in regions where warming has been small or non-existent. Second, the question whether the NE Atlantic will continue to warm as fast as it has over the past 25 years depends on the relative influence of regional processes such as the Atlantic Multidecadal Oscillation and how these change in future.

Table 5.2 Ocean climate variables grouped by type of property

Property type	Variable
Atmospheric and sea-surface	Wind
	Cloud cover
	Waves
	Sea level
Chemical and physical	Temperature
	Salinity
	pH
	Oxygen
Dynamic	Currents
	Stratification
	Turbulence
	Upwelling
	Frontal processes
Seasonal	Monsoon timing

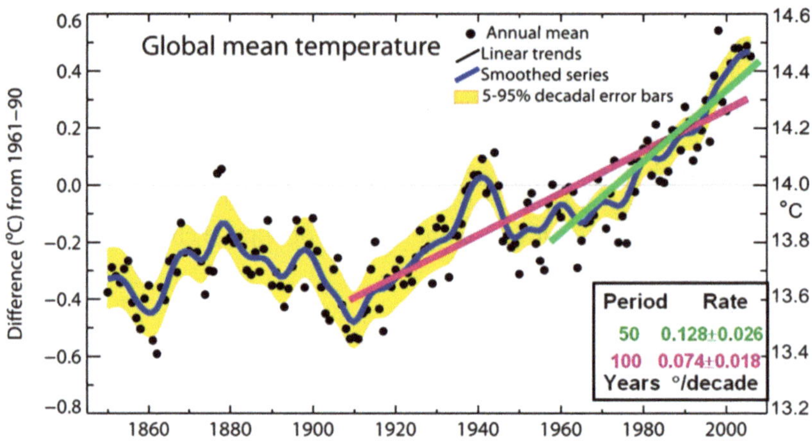

Fig. 5.1 Global mean temperature showing accelerating warming (Adapted from IPCC 2007)

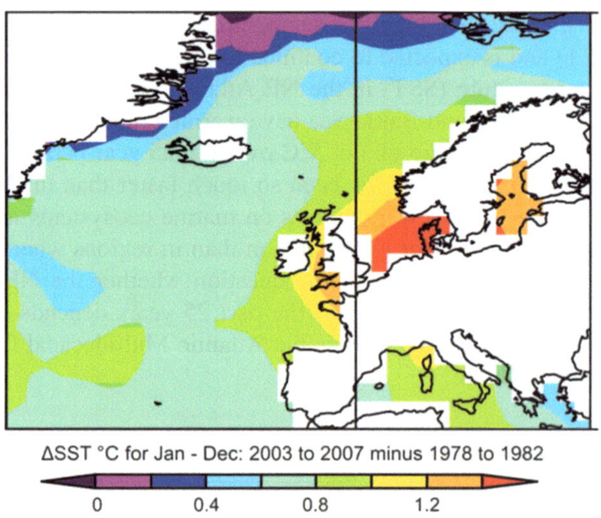

Fig. 5.2 Change in sea surface temperature (SST) between 1980 and 2005 (mean for 2003–2007 minus mean for 1978–1982) (Data are from NOAA NCDC ERSST version 2, which is an extended reconstruction of global SST, based on ICOADS (Worley et al. 2005) monthly summary trimmed group data (www.cdc.noaa.gov/))

Local and regional variability in temperature (and other climatic factors) is much greater than global variability; short-term variability (seasonal, inter-annual, decadal) is greater than climate trends. This is because local and short-term variability is smoothed out by averaging over larger spatial or temporal scales. For example, mean SST for the Baltic Sea can fluctuate by more than 2°C from year to year, which is an annual change in temperature roughly two orders of magnitude greater than the global warming trend. The seasonal variability in the Baltic Sea is also very large (mean SST varies between about 3°C in March and 18°C in July–August; Holliday et al. 2009). Since species are adapted to such large inter-annual and seasonal variability it is difficult (but not impossible) to detect the impact of relatively small global climate change. Furthermore, temperature is not the only climate variable affecting marine life; in the Baltic Sea salinity is more important in limiting the

Box 5.1: Definitions from the Glossary of the 2007 IPCC Report (WG1)

Climate in a narrow sense is usually defined as the average weather, or more rigorously, as the statistical description in terms of the mean and variability of relevant quantities over a period of time ranging from months to thousands or millions of years. The classical period for averaging these variables is 30 years, as defined by the World Meteorological Organization. The relevant quantities are most often surface variables such as temperature, precipitation and wind. Climate in a wider sense is the state, including a statistical description, of the climate system.

Climate prediction A climate prediction or climate forecast is the result of an attempt to produce an estimate of the actual evolution of the climate in the future, for example, at seasonal, inter-annual or long-term time scales. Since the future evolution of the climate system may be highly sensitive to initial conditions, such predictions are usually probabilistic in nature.

Climate projection A projection of the response of the climate system to emission or concentration scenarios of greenhouse gases and aerosols, or radiative forcing scenarios, often based upon simulations by climate models. Climate projections are distinguished from climate predictions in order to emphasize that climate projections depend upon the emission/concentration/radiative forcing scenario used, which are based on assumptions concerning, for example, future socioeconomic and technological developments that may or may not be realised and are therefore subject to substantial uncertainty.

Climate scenario A plausible and often simplified representation of the future climate, based on an internally consistent set of climatological relationships that has been constructed for explicit use in investigating the potential consequences of anthropogenic climate change, often serving as input to impact models. Projections often serve as the raw material for constructing climate scenarios, but climate scenarios usually require additional information such as about the observed current climate.

Detection and attribution Climate varies continually on all time scales. Detection of climate change is the process of demonstrating that climate has changed in some defined statistical sense, without providing a reason for that change. Attribution of causes of climate change is the process of establishing the most likely causes for the detected change with some defined level of confidence.

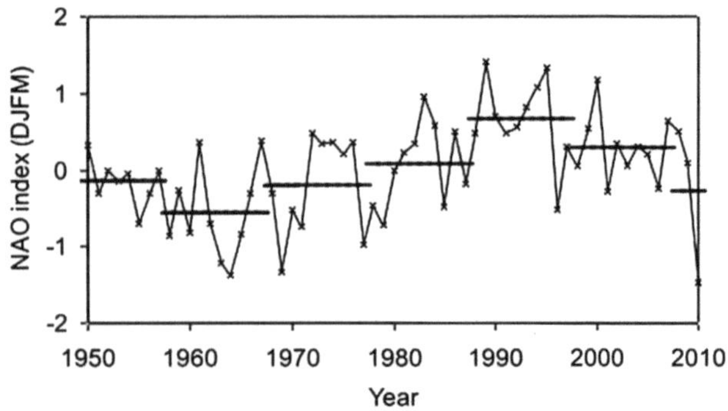

Fig. 5.3 Mean winter (DJFM) NAO values since 1950. *Horizontal lines* are decadal means (1958–1967 etc.) to match the time periods used in Fig. 5.4 (Data from www.cpc.noaa.gov/products/precip/CWlink/pna/nao_index.html)

distribution and abundance of some species, and seasonal variability in pH is far greater than in the open ocean because of the reduced alkalinity of the Baltic Sea.

The regional climate for the NE Atlantic is strongly influenced by the persistent pattern of high and low pressure known as the North Atlantic Oscillation (Hurrell and Deser 2010). This pressure gradient between Iceland (low pressure) and the Azores (high pressure) governs the strength and direction of the prevailing winds, particularly in winter (December to March), which in turn govern SST, vertical mixing and the onset of stratification during spring and early summer. A high value of the NAO index means a steep pressure gradient that drives strong westerly winds carrying warmth and moisture over Europe. A low NAO gives more northerly and easterly winds and hence colder winters for much of western Europe. The cold winter of 2009/10 was due to NAO values that were lower than any in the time series back to 1950 (Fig. 5.3), with an average matching the winters of 1963, 1964 and 1969.

During the decade 1988–1997, which had the highest NAO during the 40 year period 1958–1997, strong SW winds blew across the North Sea during February and March, affecting stratification, temperature and hence the plankton spring bloom (Fig. 5.4 from Siegismund and Schrum 2001). The increased wind strength had major consequences for plankton production and composition and for species at higher trophic levels, such as cod (Beaugrand et al. 2002, 2003). Changes in the regional windfield also affect the likelihood of major inflows of high salinity, oxygenated water into the Baltic Sea (Siegismund and Schrum 2001), but the processes are complex and not fully understood. It is likely that the NAO will become more positive due to climate change in future (Gillett et al. 2003), but here too, the processes are not well understood.

As well as affecting inter-annual variability in temperature and stratification during the plankton production season in spring and hence both the quantity and composition of phytoplankton production, the NAO also affects the biogeography of the NE Atlantic. The boundaries of the transition zone between regions of "subpolar" and "subtropical" phytoplankton blooms shift under different NAO states (Fig. 5.5). Atmospheric climate patterns such as the NAO interact dynamically with ocean circulation, in particular the subpolar gyre, to produce biogeographic shifts at all faunal levels (Hatun et al. 2009). The transition zone between warm temperate and cold temperate species has shifted northwards since the 1960s, causing fairly abrupt

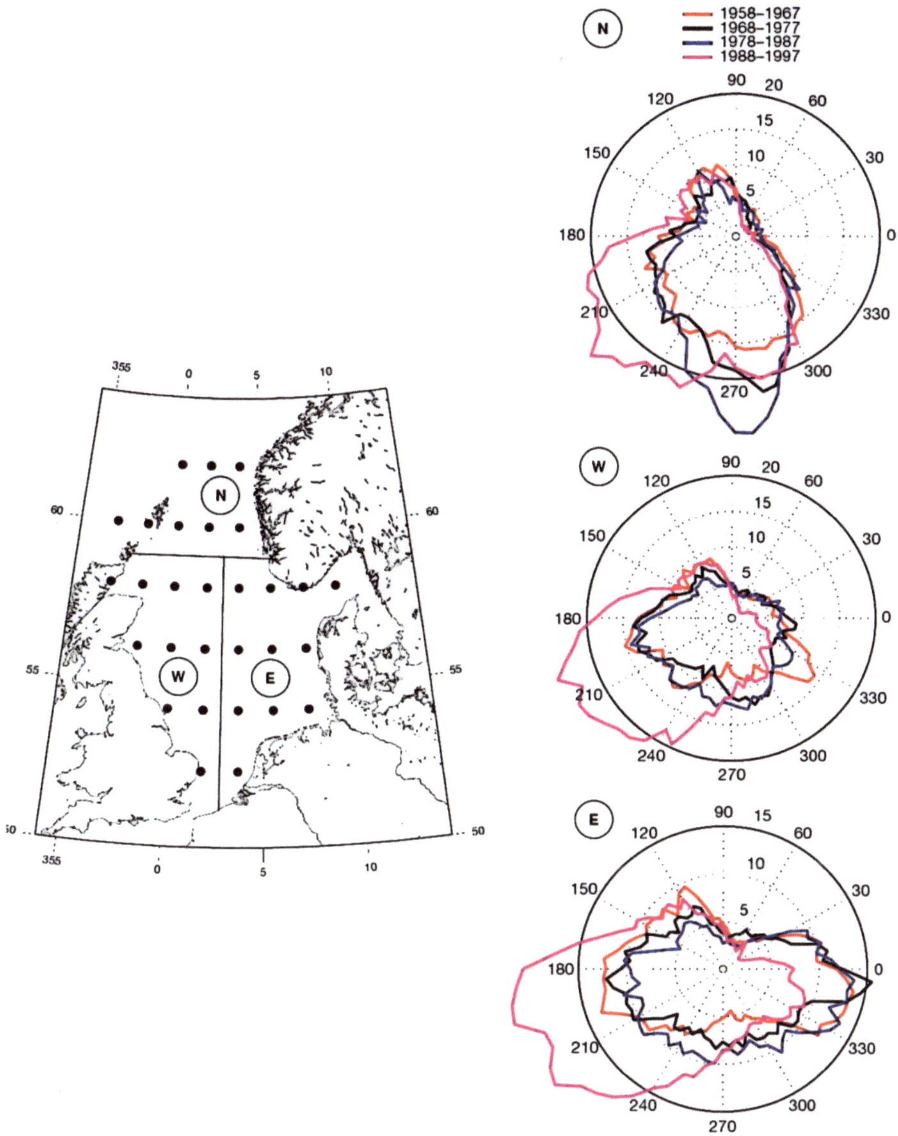

Fig. 5.4 Normalized decadal wind density functions for February–March for each of the four decades from 1958 to 1997, using NCEP re-analysis data for the grid points shown (After Siegismund and Schrum 2001). The wind speed in ms^{-1} is represented by the radial distance, with the scales shown on each *circle plot*. Note that the scale for area E is different from the other two

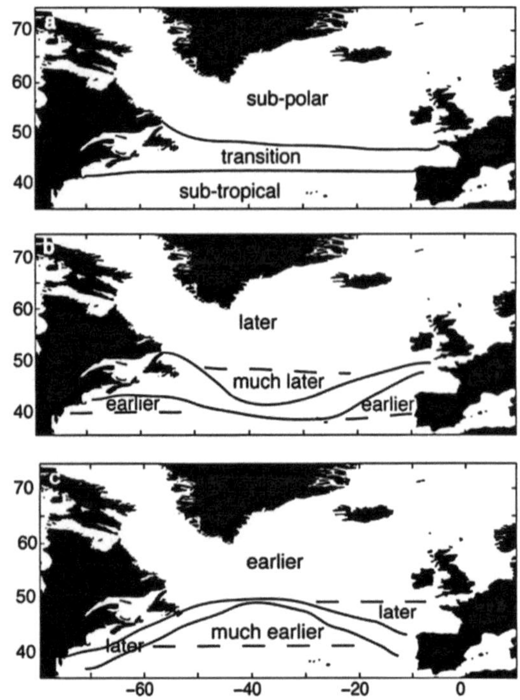

Fig. 5.5 Shift in biogeographic regions from (**a**) mean conditions to (**b**) positive NAO and (**c**) negative NAO. The *dashed line* marks the position of the transition region under mean conditions. "later" and "earlier" refers to bloom timing (From Henson et al. 2009)

changes in marine biota that can be described as "regime shifts" (Beaugrand et al. 2008).

5.3 Ocean Acidification

The oceans act as a major buffer for atmospheric gases. Approximately 30% of the additional atmospheric CO_2 that has been released since the onset of industrialisation has been absorbed by the oceans, causing the average pH to fall by 0.1 units. This may seem a small change but on the logarithmic pH scale this is equivalent to a

Box 5.2: Ocean Acidification: A Simplified Primer
In seawater, CO_2 dissolves to form carbonic acid, which dissociates to form equilibrium with hydrogen ions, bicarbonate ions and carbonate:

$$CO_2 + H_2O \leftrightarrow H_2CO_3 \leftrightarrow HCO_3^- + H^+ \leftrightarrow 2H^+ + CO_3^{2-}$$

This equilibrium is dominated by bicarbonate ($CO_2 \approx 1\%$, $CO_3^{2-} \approx 8\%$ and $HCO_3^- \approx 91\%$, of total dissolved inorganic carbon, "DIC"). Continued uptake of CO_2 by the oceans drives this equilibrium to the centre, increasing the concentration of hydrogen ions, and thereby reducing pH. This pH shift changes the equilibrium between bicarbonate and carbonate, driving that balance also toward the centre, thereby depleting the available carbonate pool.

This process increases the rate of dissolution of deposited calcium carbonate ($CaCO_3$). The rate of dissolution depends on the crystalline form of the $CaCO_3$: aragonite (found in corals and molluscs) is twice as soluble as calcite (found in echinoderms and crustaceans).

~25% increase in acidity. The chemical processes that underlie this process are well understood (Box 5.2), and consequently scenarios for future changes in ocean pH can be predicted relatively reliably from scenarios of future atmospheric CO_2 concentrations. Those scenarios project possible reductions in open ocean pH of as much as 0.4 units (equivalent to 3× more acidity) within the coming 100 years.

If realised, these changes could have substantial consequences for marine organisms, but how good is the evidence that ocean acidification is occurring? In contrast to other metrics of ocean climate change, relatively few data are available to assess whether ocean pH has changed over multi-decadal timescales. The two longest available time-series data sets come from the tropical north Atlantic and Pacific, and show clear trends of increasing dissolved CO_2 and declining pH (Fig. 5.6). The shorter "European Station for Time series in the Ocean" (ESTOC) time series from the Canary Islands in the Atlantic shows a similar pattern. In all cases the rate of pH decline with time varies with location and with depth, but is typically on the order of −0.01 to −0.02 pH units per decade at these open ocean sites. This is more rapid than

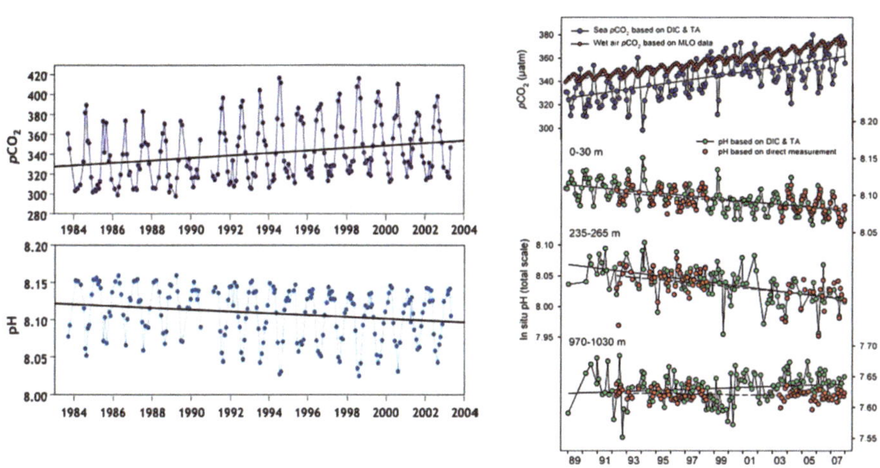

Fig. 5.6 Changes in dissolved CO_2 and pH from the Bermuda Atlantic Time Series (*left*, BATS: bats.bios.edu) and Pacific Station ALOHA (*right*, Dore et al. 2009) (Copyright 2008 National Academy of Sciences, USA)

changes over the previous century (0.1 pH units over the past 250 years, IPCC 2007) and consistent with the increasing rates of acidification that have been predicted. However, recent publications have warned that such short time-series may be unreliable for predicting longer term trends (Gruber 2009).

Closer inshore trends are less consistent. Patterns for the last 20 years in the Baltic Sea show marked reductions, no change, or even increases, in pH over time (Anderson et al. 2008). This is perhaps to be expected since the impact of climate change on terrestrial ecosystems as well as changing land-use, eutrophication, and remineralisation are all likely to have a major effect in these areas (Omstedt et al. 2009, 2010). Reductions of coastal seawater pH are also likely due to deposition of NO_x and SO_x from agriculture and fossil fuel sources and may exacerbate coastal ocean acidification (Doney et al. 2007), especially in the Baltic Sea, where the density of shipping traffic is amongst the heaviest in the world. The magnitude of these effects in the Baltic Sea is unknown.

Only one long-term time-series of observations from near-shore waters is available, from the Pacific coast of Washington State, USA (Wootton et al. 2008, but see also Thomsen et al. 2010). This shows that decadal-scale pH decline can be up to 10× more rapid than that in the open ocean, and that seasonal and diurnal pH excursions can be up to ±0.5 pH units (Wootton et al. 2008, Fig. 5.7). Again, these changes are far greater than would be expected from increasing atmospheric CO_2 concentrations, and indicate the potential importance of other processes such as interannual variation in upwelling, run-off, and community composition as well as diurnal variation in photosynthesis and respiration. Diurnal measurements in shallow bays of the Skagerrak also show fluctuations of similar magnitude (±0.2 to 0.5 pH units; Havenhand et al. 2009 and therefore it seems likely that many marine organisms are already adapted to substantial shifts in pH. The effects of reducing the global mean pH by 0.4 units on top of these fluctuations is as yet unknown.

Investigations of the effects of ocean acidification on marine organisms only became common after the issue came to a wider audience in the mid-2000s (Raven et al. 2005; Kleypas et al. 2006). There are still far more questions than answers, however one result that emerged early was that effects are variable and that even closely related species can respond very differently to the same pH change. For example, fertilization success in the sea urchin *Heliocidaris erythrogramma* was 25% lower in acidified treatments (pH = 7.7 vs. 8.1 in controls, Havenhand et al. 2008) whereas in the sister species *H. tuberculata* fertilization was unaffected by the same pH change (Williamson et al. unpublished data). Similar results have been obtained for molluscs (cf. Kurihara et al. 2007; Havenhand and Schlegel 2009; Parker et al. 2009), as well as for larvae of echinoderms (Dupont et al. 2010b). This has recently been confirmed in the first rigorous meta-analysis of ocean acidification, which found substantial – and statistically significant – variation in responses to ocean acidification among taxa within different processes (e.g. calcification, growth) as well as within processes among different taxa (e.g. growth in echinoderms, and survival in molluscs, Kroeker et al. 2010).

Calcifying species such as coccolithophores, echinoderms, and molluscs have been a focus of ocean acidification research. These species have been presumed to

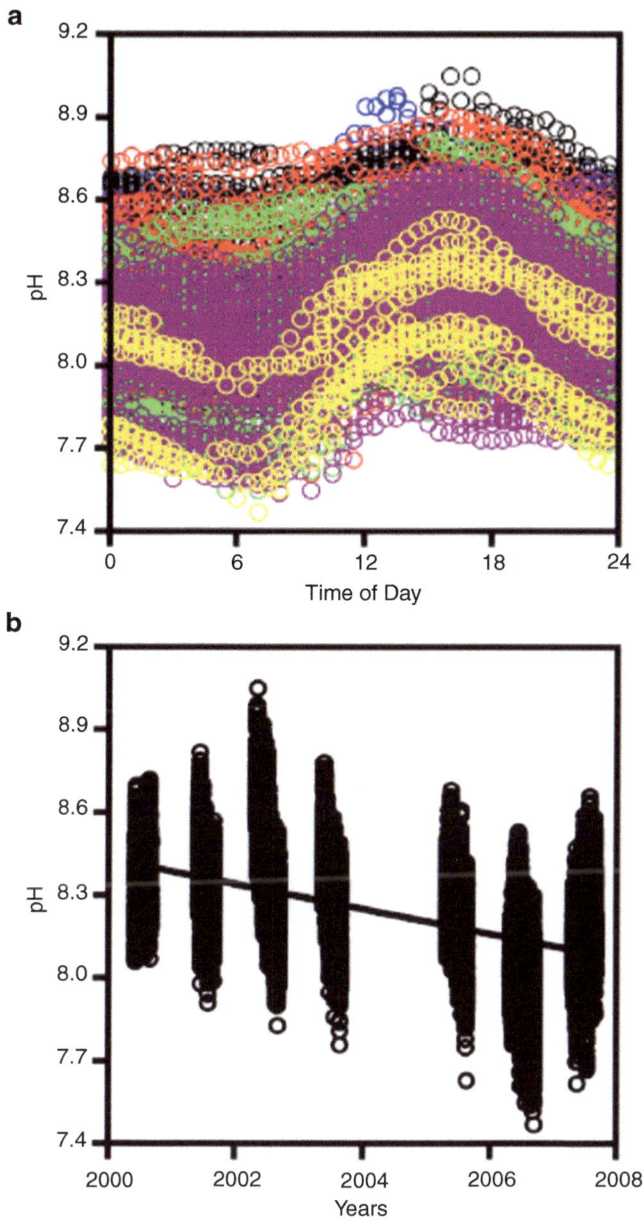

Fig. 5.7 Patterns of ocean pH through time at Tatoosh Island, WA, USA. (**a**) Daily cycle of pH arising from photosynthetic uptake of CO_2 by algal primary producers. *Colours* indicate month that the data were collected (*Blue*: April; *Black*: May; *Red*: June; *Green*: July; *Purple*: August; *Yellow*: September). (**b**) pH readings as a function of date and time taken between 2000 and 2007 (From Wootton et al. (2008). Copyright 2008 National Academy of Sciences, USA)

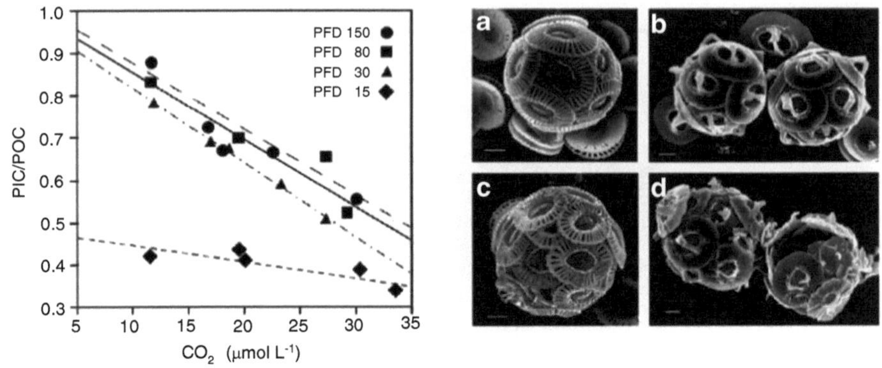

Fig. 5.8 Effects of increasing acidification (CO_2 concentration in seawater) on coccolithophores. *Left*: Decreases in the ratio of calcification to organic carbon production (PIC/POC ratio) at different levels of illumination (PFD). *Right*: Increased malformation and decreased calcification in *Emiliania huxleyi* (**a**, **c**) and *Gephyrocapsa oceanica* (**b**, **d**) in control (300 μatm CO_2, *top*) and acidified (780–850 μatm CO_2, *bottom*) seawater (After Riebesell 2004)

be more susceptible because reduced pH will affect deposition and dissolution rates of their calcium carbonate skeletons (Box 5.2, Raven et al. 2005). Early studies showed clear negative effects of ocean acidification on growth of coccolithophores (Riebesell 2004, Fig. 5.8), however subsequent work found contrary (Iglesias-Rodriguez et al. 2008) and variable (Langer et al. 2009) results, and it now seems clear that responses vary between different genetic strains of the same species (Langer et al. 2009).

Calcifying species generally respond negatively to ocean acidification, although there is much inter-specific (and inter-taxon) variation. Particularly vulnerable and ecologically important species such as corals and high-latitude pteropods have been the focus of many studies (e.g. Hoegh-Guldberg et al. 2007; Fabry et al. 2008; Comeau et al. 2009). These species are especially vulnerable because their $CaCO_3$ skeletons are laid down in the form of aragonite, rather than calcite. Aragonite is more water soluble than calcite and therefore the effects of ocean acidification are predicted to be more extreme in aragonite-forming species (coralline algae, corals, pteropods, some molluscs), than in calcite-forming species (foraminifera, coccolithophores, some macroalgae, some molluscs, crustaceans and echinoderms; note that many molluscs use mixtures of aragonite and calcite in their shells).

Tropical coral reefs typically occur in waters with a high aragonite saturation state ($\Omega_{aragonite} \geq 3.3$), which promote calcification and coral growth. Net growth of these corals only occurs above saturations states of ≈ 3; conditions that have prevailed for hundreds of thousands of years. Predictions that aragonite saturation state may fall below this level within the next 50 years imply compromised coral growth and the decline of coral reefs (Orr et al. 2005; Hoegh-Guldberg et al. 2007 and refs therein). The extent of coral reef decline in the coming decades is hotly debated, however it seems fairly certain that impacts will be measurable – and possibly catastrophic before the end of the twenty-first century.

The ecological importance of corals is easy to understand, however shelled pteropods barely feature in most marine ecological texts. Yet pteropods are ecologically important, and possibly keystone species in high-latitude pelagic communities where they can form a major component of the diet of vertebrates. They are particularly at risk because these cold polar waters are acidifiying more rapidly than the tropics. Indeed, some parts of the Arctic Ocean have already been reported to be seasonally undersaturated for aragonite ($\Omega_{aragonite} < 1$, Yamamoto-Kawai et al. 2009), and predictions suggest this pattern will be common within a decade (Steinacher et al. 2009). Recent work, however, has shown that the common pteropod *Limacina helicina* can actively precipitate its aragonite shell at $\Omega_{aragonite} < 1$ (Comeau et al. 2010), again highlighting the potential importance of variability between species in their responses to ocean acidification.

Data for calcifying species in the Baltic Sea show that they are also likely to be affected by future ocean acidification. Larvae and adults of the blue mussel, *Mytilus edulis*, and the Japanese oyster, *Crassostrea gigas*, have both been shown to decrease calcification rates as CO_2 levels increase (Gazeau et al. 2007, 2010; Kurihara et al. 2007, 2009), and it has been suggested that the absence of coccolithophores in the Baltic Sea is due to already-present undersaturation for calcite during winter months (Tyrrell et al. 2008). Calcareous echinoderms are largely absent from the Baltic proper, however, larvae of several echinoderm species in the Kattegat and Skagerrak generally show reduced growth and calcification rates in acidified conditions (though not always: see Dupont et al. 2010a).

These broad patterns of the effects on calcification are summarised clearly in the meta-analysis of Kroeker et al. (2010), which showed statistically significant variability in the effects of ocean acidification on calcifying taxa. They found a significant net negative effect of ocean acidification[1] on calcification in (tropical) corals, no significant net effect on calcareous algae, coccolithophores, molluscs and echinoderms, and a significant net positive effect on calcification in crustaceans. The meta-analysis also showed that aragonite-forming species were more negatively affected by ocean acidification than calcite-forming species, but this difference was not great (Kroeker et al. 2010).

The focus on calcifying species has been useful in rallying research on the effects of ocean acidification, however declining pH will also have a pervasive influence on other key processes such as physiology, primary production, and reproduction (especially at highly sensitive early stages of the life-history, Pechenik 1987; Raven et al. 2005), as well as broader emergent effects on whole ecosystems and the services they provide.

Here too there is a rapidly increasing literature, and again it is clear that different species respond differently. Physiological effects of ocean acidification are perhaps best summarised in the conceptual framework of Pörtner and others (e.g. Pörtner and

---

[1] Kroeker et al. (2010) restricted their analysis to studies that had used CO_2 levels and pH values equivalent to IPCC scenarios for the year 2100, i.e. $\leq$950 ppm CO_2 and a pH change of $\approx$0.4–0.5 units from present day values.

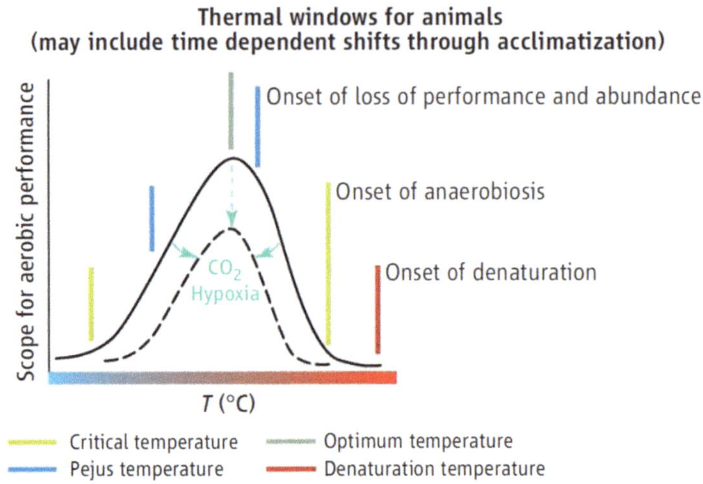

Fig. 5.9 Thermal aerobic performance windows of organisms display an optimum, limited by regions of reduced perfomance ("pejus"). Multiple stresses (here increasing CO_2) reduce both maximum performance and breadth of the thermal window (From Pörtner and Farrell 2008)

Farrell 2008). This suggests that increasing CO_2 content of the seawater (and accompanying decreases in pH) will act to reduce the overall aerobic scope of species, and particularly will narrow their window of thermal tolerance (Fig. 5.9). Several studies have now demonstrated the applicability of this concept in Baltic species (e.g. Metzger et al. 2007), although most studies have used CO_2 levels far higher than those expected within the coming century. Some authors have suggested that taxa with larger body size and more complex physiologies, such as fish and cephalopods, are likely to be less affected (Melzner et al. 2009), and in this context it is interesting that Kroeker et al. (2010) found a small positive effect of ocean acidification on growth in fish. In this context it is interesting that recent results from Baltic cod show tolerance to, and negative impacts of, ocean acidification on different early life-stages (Frommel et al. 2011 whereas metabolism and swimming of adult cod appeared unaffected by substantial reductions in pH (Melzner et al. 2009). This is broadly consistent with Pörtner and Farrell's (2008) assertion that in taxa with smaller body size and simpler physiology (e.g. early life-history stages of fish) narrowing of the thermal window by ocean acidification will probably have pervasive effects on the energy budgets and "scope for growth".

With respect to photosynthesis, the effects of ocean acidification are again variable, but generally show few trends with ocean acidification. Kroeker et al. (2010) report that no net effect was observed in seagrasses, but that calcareous algae were negatively impacted. This difference is illustrated well in the naturally acidified waters around CO_2 seep-sites in Italy (Hall-Spencer et al. 2008), where the growth of seagrasses was greater in highly acidified waters (pH $\approx$ 7.0–7.6). However this increase was strongly correlated to an absence of epiphytic calcareous algae, confounding the possible cause of the observed differences. Recent work in Baltic seagrass ecosystems also shows that eelgrasses respond positively to ocean acidification but that this effect is counteracted by increased temperature and epiphytic grazing (Eklöf et al. 2012). Once again, it seems that the effects of ocean acidification are variable (in this case in response to interactions with other components of the ecosystem).

The future effects of ocean acidification are most likely to be felt at the most sensitive, early, stages of the life-cycle, namely fertilization and larval development. Investigations of fertilization success have found negative (Havenhand et al. 2008; Parker et al. 2009) and no (Kurihara and Shirayama 2004; Havenhand and Schlegel 2009; Ericson et al. 2010) effects of ocean acidification, whereas investigations of larval stages have generally found negative effects (see Kurihara 2008 for review, but see Dupont et al. 2010a). Summarizing these data, Kroeker et al.'s (2010) meta-analysis found that the effects of ocean acidification varied markedly with life-history stage and between taxa, and that there was too much variance to allow generalised conclusions.

This illustrates the problem of how we treat negative results in ocean acidification research: traditionally these have been ignored or misinterpreted (for example by concluding that the absence of a significant result is evidence of no effect, e.g. Byrne et al. 2009). There is now emerging a clearer framework of how to deal with these data so that meta-analyses can include *all* results, not just the statistically significant ones (Havenhand et al. 2010), thereby promising a less biased view.

Will Baltic marine species survive near-future levels of ocean acidification? Clearly some species will be more affected than others, leading to changes in community structure and possibly in ecosystem functions and services. At an intra-specific level, genetic variability and plasticity in response to ocean acidification will be key determinants of whether species will be able to withstand coming changes. Increasing evidence suggests that variation and plasticity are greater in populations from environments where pH varies widely, such as the Baltic Sea, but like much of ocean acidification research, these patterns are only beginning to be investigated and we still lack answers to the most important questions.

5.4 Direct and Indirect Effects of Climate

Climate change has both direct and indirect impacts on fish and other taxa. Direct effects act on physiology and behaviour and alter growth, development, reproductive capacity, mortality, and distribution (Perry et al. 2010). Indirect effects alter the productivity, structure, and composition of the ecosystems on which species depend for food and shelter. Plankton species composition, productivity, and phenology (seasonal timing) are affected by climate, with consequences for the different life history stages of fish that depend on them (Beaugrand et al. 2003). In the Baltic Sea it has been shown that changes in distribution and abundance of commercially exploited species such as cod are due to climate effects on a competitor or predator (e.g. sprat, herring). In other areas the effect of climate may act indirectly via a parasite. Climate affects aquaculture as well as capture fisheries.

Climate is defined by long-term (30 year) averages (see Box 5.1) and the impacts of climate change can therefore be expected to be gradual, causing changes in distribution and abundance of populations as a result of slow cumulative effects on individuals. Populations of particular species may respond to climate change by shifting their distribution or by adapting to the new conditions. Climate induced distribution shifts will generally occur gradually, as a result of changes in birth and death rates at the population margins, rather than by individuals actually migrating. A simple "null model" of expected rates of distribution shift due to the anthropogenic component of climate change can be derived from global temperature patterns and expected rates of global warming. SST decreases on average by roughly 1°C every 300 km between the equator and the polar ice edge. Expected rates of global warming are in the range 0.02–0.04°C y^{-1} (IPCC 2007), therefore if species distribution shifts matches the rate of anthropogenic warming (i.e. no in-situ adaptation) then they would be expected to move poleward by about 6–12 km y^{-1}. Given the range of uncertainty in determining the distribution limits of most marine species and the variability in thermal patterns due to natural climate variability, it is likely to take several decades before such cumulative shift in species distribution can be detected. Of course there are many examples of rapid marine species distribution shifts, but these are much more likely to be due to natural climate variability or other factors than to anthropogenic warming.

The proposed "null model" does not apply to areas, such as the Baltic Sea, where other factors (e.g. salinity) govern distribution and where there may be abrupt thresholds such that a small change in salinity causes an abrupt drop in survival. The "null model" also neglects changes in climate variability and in frequency of extreme events that may result from anthropogenic climate change. These may also result in abrupt changes in survival and hence in abundance and distribution. For example extreme high temperatures in the North Sea resulted in mortality of eelpout due to reduced oxygen saturation and increased metabolic demand, with consequences for the distribution of the species. Thus although climate change is generally a gradual long-term process it may also alter the frequency of short extreme events with major lasting consequences.

Changes in distribution and adaptation lie on a continuum of spatial evolutionary processes about which we know little. Some species may track climate changes through change in distribution; others may adapt or combine both responses. Species interactions are generally ignored in climate impact predictions, but obligate species combinations may be broken apart and novel communities created.

5.4.1 Direct Effects on Growth and Recruitment: The Response of Atlantic Cod to Environmental Variability

Temperature affects rates of maturation of adult cod and rates of development and survival of eggs and larvae (ICES 2005). Rapid growth in early life stages enhances survival and increases the number of recruits per unit of spawning biomass.

This results in greater surplus production and greater resilience to fishing pressure so that cod stocks at the warm end of the species range (Celtic Sea, Irish Sea, North Sea) are able to withstand higher fishing mortalities than at the cold end of the range. However, within these warmer ecosystems cod are rarely the dominant demersal fish species and they may be vulnerable to further increase in temperature, which either exposes them to temperatures beyond their tolerance range or favours their warm tolerant competitors.

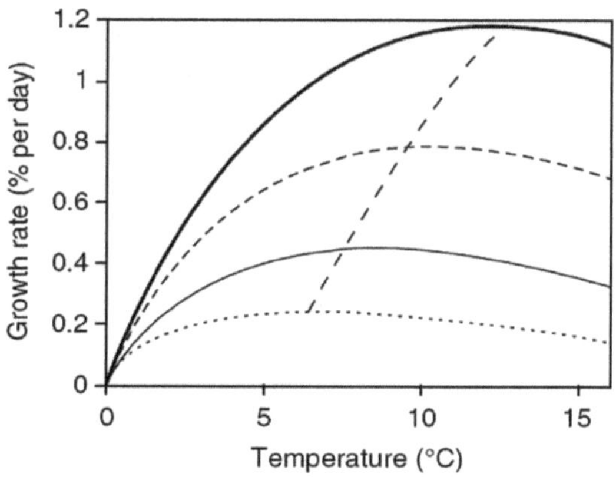

Fig. 5.10 Growth rate of four sizes of Atlantic cod (*Gadus morhua* L.) in rearing experiments at different temperatures in which they were provided unlimited food. The four fitted weights are 100 g (_____), 250 g (_ _ _), 1,000 g (____) and 5,000 g (.........). The steep *dashed line* intersects the growth curves at their maximum values, to show how the temperature for maximum growth rate declines as fish get bigger (Redrawn from Bjornsson and Steinarsson 2002)

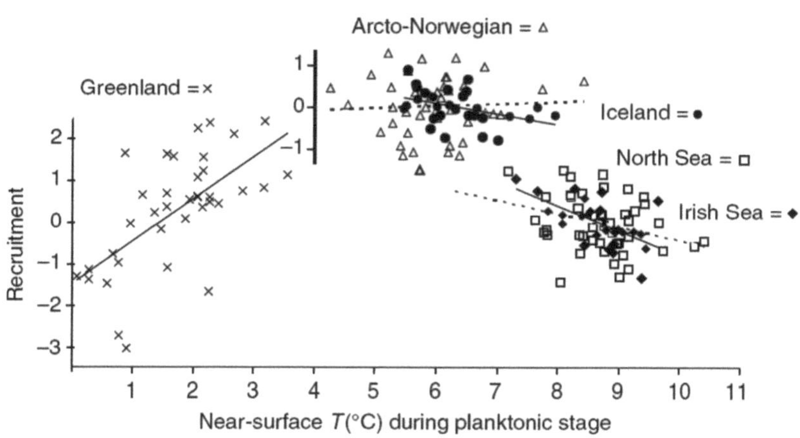

Fig. 5.11 Composite pattern of recruitment for five Atlantic cod stocks to illustrate the effect of temperature during the planktonic stage of early life on the number of recruiting fish. The scales are log *e* (number of 1-year-old fish), with the means adjusted to zero. The axes for the Arcto-Norwegian and Iceland stocks have been displaced vertically

The optimal temperature for cod growth (when food is not limiting) decreases as the fish get bigger, from about 12°C for a 100 g fish to 6°C for a 5,000 g fish (Bjornsson and Steinarsson 2002). The change in growth rate and recruitment in response to temperature is greatest at the lowest and highest temperatures within the overall tolerance range (Figs. 5.10, 5.11). The temperatures that Baltic cod experiences are generally well within their tolerance limits, except for coastal areas with extreme conditions in summer and winter.

The question of how much particular stocks of cod have become adapted to their environment is still under investigation. Genetic adaptation to extreme conditions has implications for fisheries management under climate change and special protection of warm-tolerant strains may reduce climate impacts. The fish farming industry aims to select broodstocks with good food conversion ability under specific environmental conditions, but this requires that the traits in question should be heritable (Folkvord 2005). Two examples of adaptations to extreme environmental conditions are the production of anti-freeze in Canadian stocks which experience sub-zero temperatures (Goddard and Fletcher 1994) and the production of eggs with very low density by Baltic cod, where salinity and consequently water density is much lower than in other areas (ICES 2005). The latter stock also provides one of the few cases where environmental cause and biological effect can be demonstrated all the way from physiological processes to population impact.

5.4.2 *Indirect Effects: Primary Production and Food Chain*

The rate of production in any marine ecosystem depends ultimately on the rate at which new primary production (NPP) is created by plant photosynthesis. Climate-induced changes in NPP can occur due to changes in nutrients supply and light environment, which are affected by vertical mixing, ice melt, water runoff from land, and aerial transport of nutrients (dust and sand). In the Pacific and the Atlantic oceans, nutrient supply to the upper productive layer of the ocean is declining due to reduced meridional overturning circulation, increased thermal stratification, and changes in windborne nutrients (Curry and Mauritzen 2005; McPhaden and Zhang 2002). From our present knowledge, we expect that NPP will decline in most regions due to climate change, which is an issue of concern for future global fisheries production and therefore for fisheries management (Behrenfeld et al. 2006). Polar regions will probably become more productive as the ice cover retreats, allowing greater light penetration into the ocean (Loeng et al. 2005). Satellite measurements of ocean colour over the past two decades show changes in global NPP, but with large regional differences that can be related to changes in upper ocean temperature gradients, wind stress, and atmospheric iron deposition (Lehodey et al. 2003; Gregg et al. 2003). An annual reduction in NPP of roughly 1% occurred between 1994 and 2004 (Behrenfeld et al. 2006). Paleological evidence and simulation modeling show North Atlantic plankton biomass declining by 50%

over a long time scale during periods of reduced meridional overturning circulation (Schmittner 2005).

Most fisheries exploit species that are several steps removed from primary production in the food chain. The transfer efficiency at each step is approximately 10–20%. This means that fisheries production is only a fraction of 1% of ocean primary production, and with a few notable exceptions (e.g., the Faroe marine ecosystem; Hansen et al. 2005), the relationship between NPP and fisheries production is weak. The effect of temperature on NPP is also uncertain (Sarmiento et al. 2005). In addition to quantitative changes in NPP there is evidence that climate may also result in qualitative changes that affect the length and trophic efficiency of food chains leading to fish. The qualitative changes are due to (1) changes in seasonal timing of phytoplankton production (Henson et al. 2009) (2) changes in the fraction of NPP that is produced by small and large phytoplankton groups and (3) consequent changes in the ratio of mesozooplankton production to primary production (Stock and Dunne 2009).

5.5 Distribution

As discussed earlier, the northern European continental shelf, including the Baltic Sea has undergone very rapid warming over the past 25 years and it is therefore not surprising that there have also been substantial changes in the distribution of many species of fish and plankton. However, changes in fish distribution (and in abundance) can have many causes, not all of which are climate-related or due to human activity (Table 5.3) and it is not easy to attribute observed changes to climate with confidence. The rapidly increasing public and political interest in climate change has resulted in a tendency to attribute many observed changes to climate (as anyone attending the huge range of side events at the COP15 meeting in Copenhagen in December 2009 could see), but the attribution is rarely well supported and there are often other more likely explanations.

Table 5.3 Two-way tabulation of factors that may cause changes in distribution of fish species

Causes of change	Natural	Anthropogenic
Non-climate	Competition, predation, disease, internal dynamics	Fishing, eutrophication, pollution, habitat destruction, introduction
Climate	Temperature, salinity, vertical mixing, circulation	Temperature, salinity, vertical mixing, circulation, pH

5.5.1 *Attributing Causes and Predicting Changes in Distribution*

The factors shown in Table 5.2 can interact with each other and may act at all timescales from short-term (hours to days) to long-term (years to centuries). Changes in climate factors (listed more fully in Table 5.3) can be due to natural variability or to anthropogenic effects as part of the changes brought about by rising greenhouse gas emissions. Attribution of some observed biological change to anthropogenic climate change is a two stage process, known as joint attribution or double attribution (Root et al. 2005). In the first stage change in a climate variable (usually temperature) is attributed to anthropogenic causes (GHG) (Stone et al. 2009; Allen 2003), in the second stage change in a biological state or process is attributed to the climate variable (Rosenzweig et al. 2008; ICES 2008). Part of the observed climate change since the industrial revolution can be unequivocally attributed to human emissions of GHG because GCMs that do not include these emissions fail to reproduce observed temperature trends, whereas those that include human produced GHG match the observed trends well (IPCC 2007). Biological impacts for a particular location or region are affected by the local or regional climate, which cannot be derived very reliably from global models (Drinkwater et al. 2010). Further uncertainty is added to the second stage of attribution because, unlike the IPCC class GCMs, we do not have a tested suite of biological models that can be forced with and without anthropogenic climate drivers. IPCC class GCMs produce quantitative estimates of observed and future trends in temperature caused by past GHG emissions and future GHG scenarios, but biological impact studies to date are mainly qualitative.

An attribution study for the NE Atlantic (Tasker 2008) used meta-analysis of about 300 time series from a range of marine taxa including zooplankton (83 cases), benthos (85 cases), fish (100 cases), and seabirds (20 cases). The conclusion was that about 75% of the observed biological changes in distribution, phenology and abundance are in the direction expected due to climate change and that it is highly likely that climate has had a detectable effect, however the study did not attempt to partition the climate induced change into its anthropogenic and "natural" components, except to note that the latter dominate. We need to maintain a realistic and critical view of the issues and of the difficulty of detecting climate impacts, identifying the processes and developing credible predictions (Brander 2009).

When considering climate impacts a number of distinct questions can be addressed:

- Are we interested in past effects or future effects?
- Do we only want to determine whether there has been an effect of climate or do we also want to know how big the effect has been and whether the relationship is linear or non-linear?
- Can we distinguish between the effects of climate and the effects of other factors such as fishing, habitat degradation, and pollution?

- Can we distinguish between the impact of the anthropogenic component of climate change and "natural" climate variability? Do we want to evaluate changes in abundance and productivity or only changes in distribution?

In the case of non-climate factors shown in Table 5.3, the division between natural and anthropogenic causes is fairly clear, but for climate the factors are the same in both cases and it is not easy to partition them in order to attribute a proportion of the observed changes in biota to anthropogenic climate change. The partitioning of causes shown in the table is not complete and interactions between causes should not be ignored, in particular the effect that fishing has on the sensitivity of marine systems to climate impacts.

Correct attribution is important, particularly if the observed change is adverse (e.g. decline in a fish stocks), and can be rectified by identifying and adjusting the appropriate cause (e.g. overfishing). Credible predictions of future distribution change depend on correctly identifying the processes that cause change. They can then be included in process-based models and a basic test of such models is that they should be able to hindcast observed distribution changes. If they cannot do so (and it is worth noting that very few publications that make predictions actually carry out such a test) then one can have little confidence in the predictions they generate. It is very easy to make predictions, but difficult to make predictions whose credibility can be objectively supported. One way to increase confidence in the attribution is to collect a very large body of evidence and then carry out a meta-analysis (Rosenzweig et al. 2008; ICES 2008).

5.5.2 *Distribution Shifts*

Distribution shifts are well known to the general public as well as to scientists, with new species of fish being recorded in the North Sea and Baltic Sea in recent years (Fig. 5.12). A number of papers have described shifts in distribution of fish and plankton in the North Sea and adjacent sea areas and have ascribed these to changes in fishing, wind fields, water masses and temperature (Beare et al. 2004; Brander et al. 2003; Hedger et al. 2004; Quero et al. 1998; Perry et al. 2005; Rindorf and Levy 2006). Although there is a general tendency for northward shifts over the past 30 years, this pattern is not true over longer time periods and the distribution of fish species has also deepened (Dulvy et al. 2008). Since vertical thermal gradients in European shelf seas are on average at least four orders of magnitude greater than horizontal thermal gradients, a small vertical displacement is thermally equivalent to a large horizontal displacement. An analysis of changes in distribution of North Sea cod over the past century explored the effects of fishing, temperature, winds and other environmental variables (Drinkwater et al. 2010). Distribution changes have been large, as shown by fishing surveys and commercial catches; however, despite good information on climate and other possible factors, it is not possible to choose among a number of plausible explanations (climate, fishing pressure,

MORGENAVISEN
Jyllands-Posten
Fredag den 16. januar 2004 · Uge 3
Danmarks internationale avis
www.jp.dk
JPKøbenh
I Willumsens

Dyrest og billigst

1. SEKTION, SIDE 4

Tropefisk i danske farvande

Stigende temperatur er i fuld gang med at ændre havet omkring Danmark. Tropiske fisk nærmer sig, mens torsken og muligvis også rødspætten er på vej væk.

KLIMAFORANDRINGER

1. SEKTION, SIDE 5

KLIMAFORANDRINGER

Tropiske fisk på vej nordpå

Danske fiskearter på vej væk
Nye arter i danske farvande
Torsk
Tobis
Rødspætte
Rød Mulle
Multe
Ansjos
Rosa Dory
John Dory
BREDDEGRADER
55
50
45
40
35
1992
1993
1987
1995
1988
1975
1971
1966
1963

Fig. 5.12 Report on distribution shifts from the Danish Newspaper "Jyllands Posten", 16 January 2004. The headline reads: "Tropical fish in Danish waters"

meta-population dynamics, biological interactions with prey fields). Our inability to explain such past patterns of change in a well studied area mandates caution with regard to the credibility of future projections, even if we had reliable regional climate projections, due to biological and environmental complexities.

5.6 Sensitivity of Baltic Sea Fish Populations to Climate Change

Because the Baltic Sea is only a few thousand years old and has low salinity, the diversity of its marine fauna is much lower than the adjacent North Sea. Climate change is likely to increase the input of fresh water and reduce the inflow of saline water from the North Sea, resulting in lower salinity and probably leading to reductions and losses of species that cannot tolerate these conditions.

Archaeological studies of fish-bone remains from human Stone Age settlements in Denmark provide an indication of the local fish fauna and fisheries during the Atlantic Warm Period (about 7000–3900 BC). Sea temperatures were 2–3°C higher than late twentieth century temperatures and provide an analogue of what we can expect under future warming. The fish-bone remains included species typical of warmer areas to the south of Denmark, such as anchovy, red mullet, swordfish, European sea bass and black sea bream, all of which are now returning to Danish waters.

Somewhat surprisingly, the remains of cod were also common. Recruitment of cod is expected to decline as temperatures increase, but the negative effect on the population size may have been offset by the low fishing pressure during the Stone Age. Cod populations may be able to survive the warmer temperature conditions that are likely to prevail in Danish waters during the next century, provided that the exploitation rate is kept low.

During the twentieth century, the Baltic Sea marine ecosystem was in effect subjected to a large-scale ecological engineering experiment, as humans eradicated most of the marine mammals, added nutrients and increased fishing mortalities.

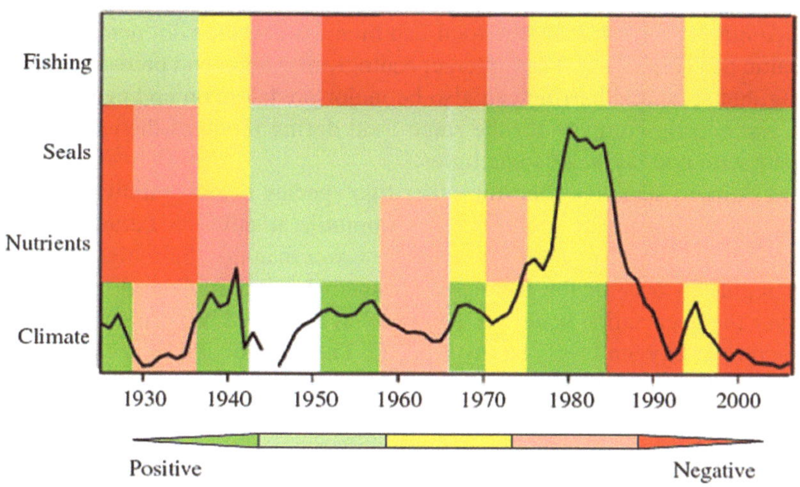

Fig. 5.13 Factors affecting spawning stock biomass of eastern Baltic cod (*Gadus morhua*) from 1925 to 2006. The *colours* represent the influence (positive–negative) of different factors on *G. morhua* biomass (From Eero et al. (2011), reproduced with permission from the Ecological Society of America)

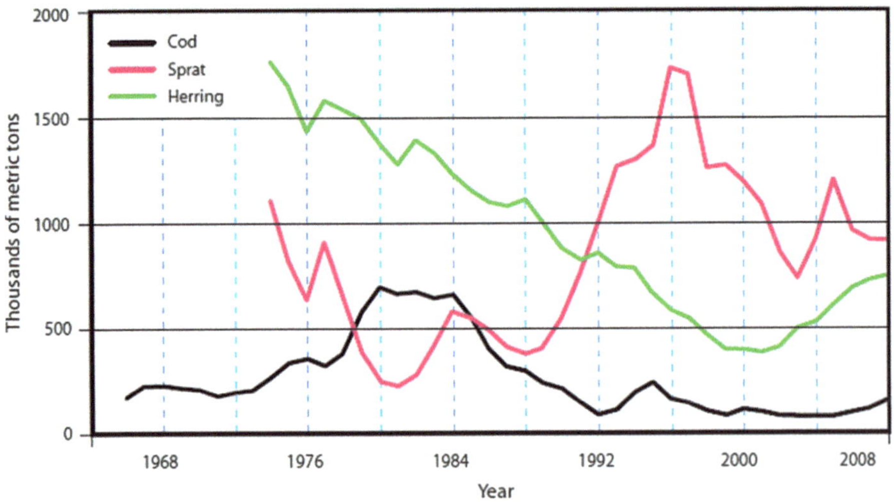

Fig. 5.14 The biomass of adult cod, sprat and herring in the Baltic Sea east of Bornholm, 1966–2008

Major variations in hydrographic conditions (sea-ice, temperature and salinity), associated with multi-decadal North Atlantic regional-scale variability, also had their effects. Analysis of long time-series of data for cod and other fish species provides an excellent opportunity to study the combined impacts of human and natural forcing on the Baltic Sea ecosystem and fisheries (Fig. 5.13).

Three species of fish, cod (*Gadus morhua*), sprat (*Sprattus sprattus*) and herring (*Clupea harengus*) dominate the total fish biomass and the commercial catches of the Baltic Sea (Fig. 5.14). They interact ecologically: adult cod predate on sprat and herring, but sprat and herring can also be major predators on cod eggs and larvae. Sprat and herring compete for the same food during much of their lives and also compete with cod larvae at some times.

Cod is more sensitive than the other two species to low salinity, particularly during reproduction. The sperm become immotile at salinities below 11 parts per thousand (psu) and the eggs sink at salinities lower than this. Salinities above 11 psu occur regularly only in the three deep basins (Bornholm, Gdansk and Gotland), however the deep saline water masses also have reduced oxygen and cod eggs do not survive at oxygen levels below 2 mL/L. The salinity and oxygen levels in the deep basins fluctuate, but have been declining in recent decades. Salinity is affected by runoff in the Baltic catchment basin and by the frequency of inflows from the Skagerrak. Oxygen is also affected by inflows and by eutrophication-related production and decomposition of organic matter. Only the Bornholm Basin has retained a volume of water with suitable salinity and oxygen levels for fertilization and survival of cod eggs and larvae in recent years.

A schematic history (Fig. 5.13) of the natural and human impacts on cod in the Baltic Sea since 1925 shows both positive effects (reduced predation by seals, some

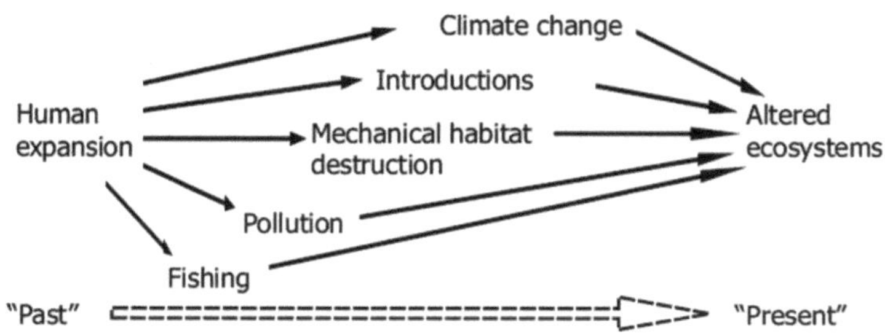

Fig. 5.15 The historic development of pressures on fisheries and marine ecosystems due to human expansion (Redrawn from Jackson et al. 2001)

productivity benefits due to increased nutrients) and negative effects (increased fishing pressure, excess nutrients causing eutrophication, adverse climate effects on salinity). The net effect of these was a remarkable increase in the biomass and catch of cod that peaked during the early 1980s, followed by a sharp decline. More restrictive fishing regulations, including lower catch quotas, longer closed seasons, larger mesh sizes and larger minimum landing sizes have been implemented to stem the decline, however the conditions that caused the decline of cod were in several ways beneficial to sprat, therefore the recovery of the cod stock is to some extent prevented by the large sprat stock.

A stochastic food web model of the Baltic Sea has been developed in order to study the interactions between the three major fish species and their environment and to provide strategic guidance on their management. It has been used to make projections of the risk that cod will decline or become extinct in the Baltic Sea under different scenarios of exploitation and climate change (Box 5.3).

5.7 Consequences of Climate Change for Biodiversity and Food Security

The aim of marine ecosystem management is to maintain healthy ecosystems and the services that they provide (Table 5.1). Climate change is one of the new, additional anthropogenic pressures on marine ecosystems (Fig. 5.15), with effects on biodiversity and productivity that are already detectable and can be expected to increase over time. Biodiversity is intrinsically valuable and therefore worthy of protection in its own right, but it also plays an important role in the resilience and productivity of marine ecosystems and therefore in maintaining food security. Another hugely important "ecosystem service" that marine ecosystems provide is their major role in global carbon flux. They sequester roughly a third of the carbon released from fossil fuels and cement around 2.3 Gt C y^{-1} in deep ocean sediments, where it may be "locked up" for geological periods of time.

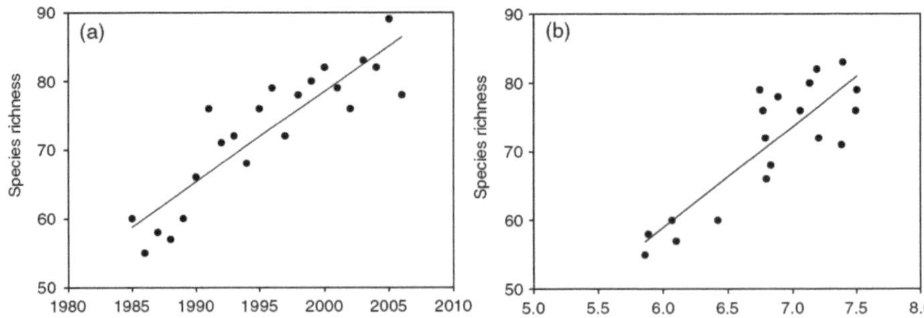

Fig. 5.16 Change in North Sea fish species richness (**a**) over time and (**b**) with temperature (From Hiddink and ter Hofstede 2008)

5.7.1 Biodiversity

Shifts in distribution, abundance and productivity have already been described and they also manifest themselves as changes in species composition and biodiversity in an area. The relationship between biodiversity and resilience (i.e. the ability of a system to recover from perturbation) is by no means simple. The North Sea and the Baltic Sea provide an interesting comparison, since the Baltic Sea has far lower species diversity, yet shows no sign of being less productive or resilient. This might suggest that there is a large amount of functional redundancy in the North Sea, but such redundancy may allow greater adaptation to changing climate conditions. However one also needs to be aware that equating "more" with "better" when thinking about biodiversity (or more specifically, species richness) is an unjustifiable simplification.

We already mentioned (Fig. 5.5) that the boundaries of the transition zone between regions of "subpolar" and "subtropical" phytoplankton blooms shift under different NAO states resulting in biogeographic shifts that can be described as "regime shifts" (Beaugrand et al. 2008). Not surprisingly, given that species richness tends to increase towards the equator, these shifts have resulted in increased species richness of plankton and fish in the North Sea during recent decades (Fig. 5.16 and Hatun et al. 2009).

5.7.2 Food Security

Food security in relation to fish depends in part on the continued health and productivity of marine ecosystems, but also on changes in income and livelihood of communities that depend on fish (Schmidhuber and Tubiello 2007; Allison et al. 2009; Brander 2007). Predictions of future NPP can only be made with very large margins of error (Sarmiento et al. 2005) and adding the trophic steps leading to fisheries production adds to these error margins. Chapter 5 of the 2007 IPCC report

(Easterling et al. 2007) concluded with medium confidence that "regional changes in the distribution and production of particular fish species are expected due to continued warming, with adverse effects projected for aquaculture and fisheries."

The adverse effects on aquaculture arise from (1) stress due to increased temperature and oxygen demand and decreased pH, (2) uncertain future water supply, (3) extreme weather events, (4) increased frequency of diseases and toxic events, (5) sea level rise and conflict of interest with coastal defenses, and (6) an uncertain future supply of fishmeal and oils from capture fisheries. Aquaculture and capture fisheries may also benefit from the effects of warming, through increased growth rates and food conversion efficiencies, longer growing season, range expansion, and the use of new areas as a result of decrease in ice cover.

Fish stocks have over time experienced a number of anthropogenic stresses that act progressively and synergistically to cause changes in distribution and abundance (Fig. 5.12). Fishing was the earliest anthropogenic pressure on fish stocks and marine ecosystems, beginning hundreds or even thousands of years ago (Jackson et al. 2001; Ojaveer and MacKenzie 2007). Anthropogenic climate change, whose impact has been detected over the past few decades, is the most recent. Management of fisheries

Box 5.3: Exploring the Effects of Climate on Cod Within the Baltic Sea Marine Ecosystem Using a Coupled Physical-Biological Model

The Baltic Sea provides an instructive regional example of the insight to be gained from effective linking of climactic and biological models while also revealing the limitations of present models. Climate projections for the Baltic Sea in the twenty-first century were based on an assessment using dynamic and statistical downscaling from GCMs (BACC 2008). A stochastic food web model (Lindegren et al. 2009) was used to quantify the interactions between the three major fish species in the Baltic Sea (cod, sprat and herring) as well as their prey, major environmental drivers and fishing pressure. Salinity plays a greater role than temperature in the biological response of cod in the Baltic Sea but the projected changes in salinity differ considerably, depending upon which global model is used to force the regional scenarios (Meier et al. 2006). A significant decrease in salinity (outside the present day range of climatic variability) is found only for the runs forced by one GCM (ECHAM4), which is also the only GCM showing statistically significant changes in wind fields in this region. The pattern and strength of wind forcing and the magnitude of precipitation are critical for Baltic Sea salinity and are not represented consistently or in detail in GCMs. The likelihood that cod will no longer be able to reproduce in the Baltic Sea depends critically on whether and by how much the salinity decreases. The stochastic food web model provides a valuable insight into a fisheries management strategy that may prevent cod biomass from dropping below the limit reference value as salinity declines (Lindegren et al. 2010), but the likelihood of such a salinity decline cannot be quantified from current climate models.

and of marine ecosystems has not yet succeeded in dealing adequately with the old pressures, and some of them, particularly overfishing, are of greater immediate concern than the effects of climate change (Beddington et al. 2007). Nevertheless, climate change over the coming decades to centuries will have progressively greater impacts on marine ecosystems and fisheries. Anticipating and adapting to such changes will help to minimize the disruption to marine ecosystems and to human food supplies. This requires models that link from IPCC class global circulation models (GCMs) to regional climate and regional biology (Stock et al. 2011).

References

Allen MR (2003) Liability for climate change; will it ever be possible to sue anyone for damaging the climate? Nature 421:892

Allison EH, Perry AL, Badjeck M-C, Adger WN, Brown K, Conway D, Hall AS, Pilling GM, Reynolds JD, Andrew NL, Dulvy NK (2009) Vulnerability of national economies to the impacts of climate change on fisheries. Fish Fish 10:173–196

Anderson P, Håkansson B, Håkansson J, Sahlsten E, Havenhand JN, Thorndyke M, Dupont S (2008) Marine acidification – on effects and monitoring of marine acidification in the seas surrounding Sweden. SMHI Oceanography 92, SMHI, Norrköping, Sweden, 61 pp

BACC Author Team (2008) Assessment of climate change for the Baltic sea basin. Regional climate series. Springer, Berlin/Heidelberg, 473 pp

Beare DJ, Burns F, Greig A, Jones EG, Peach K, Kienzle M, McKenzie E, Reid DG (2004) Long-term increases in prevalence of North sea fishes having southern biogeographic affinities. Mar Ecol Prog Ser 284:269–278

Beaugrand G, Reid PC, Ibanez F, Lindley JA, Edwards M (2002) Reorganization of North Atlantic marine copepod biodiversity and climate. Science 296:1692–1694

Beaugrand G, Brander KM, Lindley JA, Souissi S, Reid PC (2003) Plankton effect on cod recruitment in the North sea. Nature 426:661–664

Beaugrand G, Edwards M, Brander KM, Luczak C, Ibanez F (2008) Causes and projections of abrupt climate-driven ecosystem shifts in the North Atlantic. Ecol Lett 11:1–12

Beddington JR, Agnew DJ, Clark CW (2007) Current problems in the management of marine fisheries. Science 316:1713–1716

Behrenfeld MJ, O'Malley RTO, Siegel DA, McClain CR, Sarmiento JL, Feldman GC, Milligan AJ, Falkowski PG, Letelier RM, Boss ES (2006) Climate-driven trends in contemporary ocean productivity. Nature 444:752–755

Bjornsson B, Steinarsson A (2002) The food-unlimited growth rate of Atlantic cod (*Gadus morhua*). Can J Fish Aquat Sci 59:494–502

Brander KM (2007) Global fish production and climate change. Proc Natl Acad Sci 104:19709–19714

Brander KM (2009) Impacts of climate change on marine ecosystems and fisheries. J Mar Biol Assoc India 51:1–13

Brander KM, Blom G, Borges MF, Erzini K, Henderson G, MacKenzie BR, Mendes H, Ribeiro J, Santos AMP, Toresen R (2003) Changes in fish distribution in the eastern North Atlantic: are we seeing a coherent response to changing temperature? ICES Mar Sci Symp 219:261–270

Byrne M, Ho M, Selvakumaraswamy P, Nguyen HD, Dworjanyn SA, Davis AR (2009) Temperature, but not pH, compromises sea urchin fertilization and early development under near-future climate change scenarios. Proc R Soc B-Biol Sci 276:1883–1888

Comeau S, Gorsky G, Jeffree R, Teyssie JL, Gattuso JP (2009) Impact of ocean acidification on a key Arctic pelagic mollusc (*Limacina helicina*). Biogeosciences 6:1877–1882

Comeau S, Jeffree R, Teyssie JL, Gattuso JP (2010) Response of the Arctic Pteropod *Limacina helicina* to projected future environmental conditions. Plos One 5

Curry R, Mauritzen C (2005) Dilution of the northern North Atlantic Ocean in recent decades. Science 308:1772–1774

Doney SC, Mahowald N, Lima I, Feely RA, Mackenzie FT, Lamarque J-F, Rasch PJ (2007) Impact of anthopogenic atmospheric nitrogen and sulfur deposition on ocean acidification and the organic carbon system. Proc Natl Acad Sci USA 104:14580–14585

Dore JE, Lukas R, Sadler DW, Church MJ, Karl DM (2009) Physical and biogeochemical modulation of ocean acidification in the central North Pacific. Proc Natl Acad Sci USA 106:12235–12240

Drinkwater K, Schrum C, Brander KM (eds) (2010) Cod and future climate change. ICES cooperative research report no. 305, 88 pp

Dulvy NK, Rogers SI, Jennings S, Stelzenmuller V, Dye S, Skjoldal H-R (2008) Climate change and deepening of the North sea fish assemblage: a biotic indicator of warming seas. J Appl Ecol 45:1029–1039

Dupont S, Lundve B, Thorndyke M (2010a) Near future ocean acidification increases growth rate of the Lecithotrophic Larvae and Juveniles of the sea star *Crossaster papposus*. J Exp Zool B Mol Dev Evol 314B:382–389

Dupont S, Ortega-Martinez O, Thorndyke M (2010b) Impact of near-future ocean acidification on echinoderms. Ecotoxicology 19:449–462

Easterling WE, Aggrawal PK, Batima P, Brander KM, Erda L, Howden SM, Kirilenko A, Morton J, Soussana J-F, Schmidhuber J, Tubiello FN (2007) Food, fibre and forest products. In: Parry ML, Canziani OF, Palutikof JP, van der Linden PJ, Hanson CE (eds) Climate change 2007: impacts, adaptation and vulnerability. Contribution of working group II to the fourth assessment report of the intergovernmental panel on climate change. Cambridge University Press, Cambridge, pp 273–313

Eero M, MacKenzie BR, Köster FW, Gislason H (2011) Multi-decadal responses of a cod (*Gadus morhua*) population to human-induced trophic changes, exploitation and climate variability. Ecol Appl 21:214–226

Eklöf J, Alsterberg C, Havenhand C, Sundbäck K, Wood H, Gamfeldt L (2012) Experimental climate change weakens the insurance effect of biodiversity, submitted to Ecology Letters

Ericson JA, Lamare MD, Morely SA, Barker MF (2010) The response of two ecologically important Antarctic invertebrates (*Sterechinus neumayeri* and *Parborlasia corrugatus*) to reduced seawater pH: effects on fertilisation and embryonic development. Mar Biol 157:2689–2702. doi:10.1007/s00227-010-1529-y

Fabry VJ, Seibel BA, Feely RA, Orr JC (2008) Impacts of ocean acidification on marine fauna and ecosystem processes. ICES J Mar Sci 65:414–432

Folkvord A (2005) Comparison of size-at-age of larval Atlantic cod (*Gadus morhua*) from different populations based on size- and temperature-dependent growth models. Can J Fish Aquat Sci 62:1037–1052

Frommel AY, Maneja R, Lowe D, Malzahn AM, Geffen AJ, Folkvord A, Piatkowski U, Clemmesen C. (2011) Ocean acidification effects on larvae of a commercially important fish pecies, Atlantic cod (*Gadhus morhua*). Nature Climate Change 2: 42–46.

Gazeau F, Quiblier C, Jansen CM, Gattuso JP, Middelburg JJ, Heip CHR (2007) Impact of elevated CO_2 on shellfish calcification. Geophys Res Lett 34(7):L07603

Gazeau F, Gattuso JP, Dawber C, Pronker AE, Peene F, Peene J, Heip CHR, Middelburg JJ (2010) Effect of ocean acidification on the early life stages of the blue mussel *Mytilus edulis*. Biogeosciences 7:2051–2060

Gillett NP, Graf HF, Osborn T (2003) Climate change and the North Atlantic oscillation. In: Hurrell JW, Kushnir Y, Ottersen G, Visbeck M (eds) The North Atlantic oscillation: climatic significance and environmental impact, vol 134. American Geophysical Union, Washington, pp 193–209

Goddard SV, Fletcher GL (1994) Antifreeze proteins: their role in cod survival and distribution from egg to adult. ICES Mar Sci Symp 198:676–683
Gregg WW, Conkright ME, Ginoux P, O'Reilly JE, Casey NW (2003) Ocean primary production and climate: global decadal changes. Geophys Res Lett 30:1809
Gruber N (2009) Carbon cycle: fickle trends in the ocean. Nature 458:155–156
Hall-Spencer JM, Rodolfo-Metalpa R, Martin S, Ransome E, Fine M, Turner SM, Rowley SJ, Tedesco D, Buia MC (2008) Volcanic carbon dioxide vents show ecosystem effects of ocean acidification. Nature 454:96–99
Hansen B, Eliasen SK, Gaard E, Larsen KMH (2005) Climatic effects on plankton and productivity on the Faroe shelf. ICES J Mar Sci: J Conseil 62:1224–1232
Hatun H, Payne MR, Beaugrand G, Reid PC, Sando H, Drange H, Hansen B, Jacobsen JA, Bloch D (2009) Large bio-geographical shifts in the north-eastern Atlantic ocean: from the subpolar gyre, via plankton, to blue whiting and pilot whales. Prog Oceanogr 80:149–162
Havenhand JN, Schlegel P (2009) Near-future levels of ocean acidification do not affect sperm motility and fertilization kinetics in the oyster *Crassostrea gigas*. Biogeosciences 6:3009–3015
Havenhand JN, Buttler FR, Thorndyke MC, Williamson JE (2008) Near-future levels of ocean acidification reduce fertilization success in a sea urchin. Curr Biol 18:R651–R652
Havenhand JN, Dupont S, Quinn GP (2010) Designing ocean acidification experiments to maximise inference. In: Riebesell U, Fabry VJ, Hansson L, Gattuso JP (eds) Guide to best practices for ocean acidification research and data reporting. Publications Office of the European Union Pages, Luxembourg, pp 67–80
Hedger R, McKenzie E, Heath M, Wright P, Scott B, Gallego A, Bridson J (2004) Analysis of the spatial distributions of mature cod (*Gadus morhua*) and haddock (*Melanogrammus aeglefinus*) abundance in the North sea (1980–1999) using generalised additive models. Fish Res 70:17–25
Henson SA, Dunne JP, Sarmiento JL (2009) Decadal variability in North Atlantic phytoplankton blooms. J Geophys Res 114:C04013
Hiddink JG, ter Hofstede R (2008) Climate induced increases in species richness of marine fishes. Glob Chang Biol 14:453–460
Hoegh-Guldberg O, Mumby PJ, Hooten AJ, Steneck RS, Greenfield P, Gomez E, Harvell CD, Sale PF, Edwards AJ, Caldeira K, Knowlton N, Eakin CM, Iglesias-Prieto R, Muthiga N, Bradbury RH, Dubi A, Hatziolos ME (2007) Coral reefs under rapid climate change and ocean acidification. Science 318:1737–1742
Holliday NP, Hughes SL, Beszczynska-Möller A (2009) ICES report on ocean climate 2008. ICES cooperative research report 298, 66 pp
Hurrell JW, Deser C (2010) North Atlantic climate variability: the role of the North Atlantic oscillation. J Mar Syst 79:230
ICES (2005) In: Brander KM (ed) Spawning and life history information for North Atlantic cod stocks. ICES cooperative research report 274
ICES (2008) In: Tasker ML (ed) The effect of climate change on the distribution and abundance of marine species in the OSPAR Maritime Area. ICES cooperative research report no. 293, 45 pp
Iglesias-Rodriguez MD, Halloran PR, Rickaby REM, Hall IR, Colmenero-Hidalgo E, Gittins JR, Green DRH, Tyrrell T, Gibbs SJ, von Dassow P, Rehm E, Armbrust EV, Boessenkool KP (2008) Phytoplankton calcification in a high-CO_2 world. Science 320:336–340
IPCC (2007) Climate change 2007: the physical basis. Contribution of working group I to the fourth assessment report of the Intergovernmental Panel on Climate Change (IPCC), 996 pp
Jackson JBC et al (2001) Historical overfishing and the recent collapse of coastal ecosystems. Science 293:629–637
Kleypas JA, Feely RA, Fabry VJ, Langodon C, Sabine CL, Robbins LL (2006) Impacts of ocean acidification on coral reefs and other marine calcifiers: a guide for future research. NSF, NOAA, USGS, Contribution No. 2897 from NOAA/Pacific marine environmental laboratory
Kroeker KJ, Kordas RL, Crim RN, Singh GG (2010) Meta-analysis reveals negative yet variable effects of ocean acidification on marine organisms. Ecol Lett. doi:10.1111/j.1461-0248.2010.01518.x
Kurihara H (2008) Effects of CO_2-driven ocean acidification on the early developmental stages of invertebrates. Mar Ecol Prog Ser 373:275–284

Kurihara H, Shirayama Y (2004) Effects of increased atmospheric CO_2 on sea urchin early development. Mar Ecol Prog Ser 274:161–169
Kurihara H, Kato S, Ishimatsu A (2007) Effects of increased seawater pCO_2 on early development of the oyster *Crassostrea gigas*. Aquat Biol 1:91–98
Kurihara H, Asai T, Kato S, Ishimatsu A (2009) Effects of elevated pCO(2) on early development in the mussel *Mytilus galloprovincialis*. Aquat Biol 4:225–233
Langer G, Nehrke G, Probert I, Ly J, Ziveri P (2009) Strain-specific responses of *Emiliania huxleyi* to changing seawater carbonate chemistry. Biogeosciences 6:2637–2646
Lehodey P, Chai F, Hampton J (2003) Modelling climate-related variability of tuna populations from a coupled ocean biogeochemical-populations dynamics model. Fish Oceanogr 12:483–494
Lindegren A, Mollmann C, Nielsen A, Stenseth NC (2009) Preventing the collapse of the Baltic cod stock through an ecosystem-based management approach. Proc Natl Acad Sci 106:14722–14727
Lindegren M, Möllmann C, Nielsen A, Brander K, MacKenzie B, Stenseth NC (2010) Ecological forecasting under climate change: the case of Baltic cod. Phil Trans R Soc Lond B
Loeng H et al (2005) Marine systems. In: Berner J, Symon C, Arris L et al (eds) Arctic climate impacts assessment. Cambridge University Press, Cambridge, UK, pp 453–538
McPhaden MJ, Zhang D (2002) Slowdown of the meridional overturning circulation in the upper Pacific ocean. Nature 415:603–608
Meier HEM, Kjellström E, Graham LP (2006) Estimating uncertainties of projected Baltic Sea salinity in the late 21st century. Geophys Res Lett 33:L15705. doi:10.1029/2006GL026488
Melzner F, Gutowska MA, Langenbuch M, Dupont S, Lucassen M, Thorndyke MC, Bleich M, Portner HO (2009) Physiological basis for high CO_2 tolerance in marine ectothermic animals: pre-adaptation through lifestyle and ontogeny? Biogeosciences 6:2313–2331
Metzger R, Sartoris FJ, Langenbuch M, Portner HO (2007) Influence of elevated CO_2 concentrations on thermal tolerance of the edible crab *Cancer pagurus*. J Therm Biol 32:144–151
Millennium Ecosystem Assessment (2005) Ecosystems and human well-being: synthesis. Island Press, Washington, DC, pp 71–83, Chapter 3
Ojaveer H, MacKenzie BR (2007) Historical development of fisheries in northern Europe—reconstructing chronology of interactions between nature and man. Fish Res 87:102–105
Omstedt A, Gustafsson E, Wesslander K (2009) Modelling the uptake and release of carbon dioxide in the Baltic Sea surface water. Cont Shelf Res 29:870–885
Omstedt A, Edman M, Anderson LG, Laudon H (2010) Factors influencing the acid-base (pH) balance in the Baltic Sea: a sensitivity analysis. Tellus B Chem Phys Meteorol 62:280–295
Orr JC, Fabry VJ, Aumont O, Bopp L, Doney SC, Feely RA, Gnanadesikan A, Gruber N, Ishida A, Joos F, Key RM, Lindsay K, Maier-Reimer E, Matear R, Monfray P, Mouchet A, Najjar RG, Plattner GK, Rodgers KB, Sabine CL, Sarmiento JL, Schlitzer R, Slater RD, Totterdell IJ, Weirig MF, Yamanaka Y, Yool A (2005) Anthropogenic ocean acidification over the twenty-first century and its impact on calcifying organisms. Nature 437:681–686
Parker LM, Ross PM, O'Connor WA (2009) The effect of ocean acidification and temperature on the fertilization and embryonic development of the Sydney rock oyster *Saccostrea glomerata* (Gould 1850). Glob Chang Biol 15:2123–2136
Pechenik JA (1987) Environmental influences on larval survival and development. In: Geise AC, Pearce JS, Pearce VB (eds) Reproduction of marine invertebrates, vol 9. Blackwell Scientific Publications, Palo Alto, pp 551–608
Perry AL, Low PJ, Ellis JR, Reynolds JD (2005) Climate change and distribution shifts in marine fishes. Science 308:1912–1915
Perry RI, Cury P, Brander K, Jennings S, Möllmann C, Planque B (2010) Sensitivity of marine systems to climate and fishing: concepts, issues and management responses. J Mar Syst 79:427–435
Pörtner HO, Farrell AP (2008) Physiology and climate change. Science 322:690–692
Quero J-C, Du Buit M-H, Vayne J-J (1998) Les observations de poissons tropicaux et le rechauffement des eaux dans l'Atlantique europeen. Oceanolog Acta 21:345–351

Raven J, Caldeira K, Elderfield H, Hoegh-Guldberg O, Liss P, Riebesell U, Shepherd J, Turley C, Watson A (2005) Ocean acidification due to increasing atmospheric carbon dioxide. The Royal Society, London
Riebesell U (2004) Effects of CO_2 enrichment on marine phytoplankton. J Oceanogr 60:719–729
Rindorf A, Levy P (2006) Warm, windy winters drive cod north and homing of spawners keeps them there. J Appl Ecol 43:445–453
Root TL, MacMynowski P, Mastandrea MD, Schneider SH (2005) Human-modified temperatures induce species changes: joint attribution. Proc Natl Acad Sci 102:7465–7469
Rosenzweig C, Karoly D, Vicarelli M, Neofotis P, Wu Q, Casassa G, Menzel A, Root TL, Estrella N, Seguin B, Tryjanowski P, Liu C, Rawlins S, Imeson A (2008) Attributing physical and biological impacts to anthropogenic climate change. Nature 453:353–357
Sarmiento JL, Slater R, Barber R, Bopp L, Doney SC, Hirst AC, Kleypas J, Matear R, Mikolajewicz U, Monfray P, Orr J, Soldatov V, Spall SA, Stouffer R (2005) Response of ocean ecosystems to climate warming. Global Biogeochemical Cycles 18
Schmidhuber J, Tubiello FN (2007) Climate change and food security special feature: global food security under climate change. Proc Natl Acad Sci 104:19703–19708
Schmittner A (2005) Decline of the marine ecosystem caused by a reduction in the Atlantic overturning circulation. Nature 434:628–633
Siegismund F, Schrum C (2001) Decadal changes in the wind forcing over the North Sea. Clim Res 18:39–45
Steinacher M, Joos F, Frolicher TL, Plattner GK, Doney SC (2009) Imminent ocean acidification in the Arctic projected with the NCAR global coupled carbon cycle-climate model. Biogeosciences 6:515–533
Stock CA and Dunne J (2009) Controls on the ratio of mesozooplankton production to primary production in marine ecosystems Deep-Sea Research I 57: 95–112
Stock CA, Alexander MA, Bond NA, Brander K, Cheung WWL, Curchitser EN, Delworth TL, Dunne JP, Griffies SM, Haltuch MA, Hare JA, Hollowed AB, Lehodey P, Levin SA, Link JS, Rose KA, Rykaczewski RR, Sarmiento JL, Stouffer RJ, Schwing FB, Vecchi GA, Werner FE (2011) On the use of IPCC-class models to assess the impact of climate on living marine resources. Prog Oceanogr 88:1–27
Stone DA, Allen MR, Stott PA, Pall P, Min S, Nozawa T, Yukimoto S (2009) The detection and attribution of human influence on climate. Annu Rev Environ Resour 34:1–16
Tasker ML (2008) The effect of climate change on the distribution and abundance of marine species in the OSPAR maritime area. ICES cooperative research report no. 293, 49 pp
Thomsen J, Gutowska MA, Saphörster J, Heinemann A, Trübenbach K, Fietzke J, Hiebenthal C, Eisenhauer A, Körtzinger A, Wahl M, Melzner F (2010) Calcifying invertebrates succeed in a naturally CO_2 enriched coastal habitat but are threatened by high levels of future acidification. Biogeosci Discuss 7:5119–5156
Tyrrell T, Schneider B, Charalampopoulou A, Riebesell U (2008) Coccolithophores and calcite saturation state in the Baltic and Black seas. Biogeosciences 5:485–494
Wootton JT, Pfister CA, Forester JD (2008) Dynamic patterns and ecological impacts of declining ocean pH in a high-resolution multi-year dataset. Proc Natl Acad Sci USA 105:18848–18853
Worley SJ et al (2005) ICOADS release 2.1 data and products. Int J Climatol 25:823–842
Yamamoto-Kawai M, McLaughlin FA, Carmack EC, Nishino S, Shimada K (2009) Aragonite undersaturation in the Arctic Ocean: effects of ocean acidification and sea ice melt. Science 326:1098–1100

Chapter 6
Towards Integrated Ecosystem Assessments (IEAs) of the Baltic Sea: Investigating Ecosystem State and Historical Development

Rabea Diekmann, Saskia Otto, and Christian Möllmann

6.1 Introduction

Managing marine ecosystem resources like fish is traditionally focused on single target species. However, ecosystem approaches to the management of the marine environment (EAM) now evolved to a central paradigm underlying living marine resource policy (Rosenberg and McLeod 2005; McLeod and Leslie 2009; Levin et al. 2009). EAM can be sectorial like the Ecosystem-based Fisheries Management (e.g. Hall and Mainprize 2004), or cross-sectorial also accounting for social–ecological interactions (McLeod and Leslie 2009).

The importance of the EAM for meeting the challenges of a sustainable cross-sectorial use of marine ecosystems and additionally accounting for conservation aspects is now widely accepted but it remains difficult to put the principles into practice. Various recent publications map out strategies how to implement the EAM. They usually describe iterative processes, involving a number of interdependent steps in a hierarchical system (e.g. de la Mare 2005; Crowder et al. 2008). A process that aims at making the EAM operational is the so-called Integrated Ecosystem Assessment (IEA), defined as "a formal synthesis and quantitative analysis of information on relevant natural and socioeconomic factors, in relation to specified ecosystem management objectives" (Levin et al. 2009). In IEA, an integrated scientific understanding is developed and builds the basis for management decisions. The IEA process developed by Levin et al. (2009) and expanded by Tallis et al. (2010) involves seven integrative steps: (1) scoping (identifying goals and threats), (2) defining indicators and measures, (3) setting thresholds, (4) performing risk

R. Diekmann (✉) • S. Otto • C. Möllmann
Institute for Hydrobiology and Fisheries Science, University of Hamburg, Große Elbstrasse 133, D-22767 Hamburg, Germany
e-mail: rabea.diekmann@uni-hamburg.de; saskia.otto@uni-hamburg.de; christian.moellmann@uni-hamburg.de

M. Reckermann et al. (eds.), *Climate Impacts on the Baltic Sea: From Science to Policy*, Springer Earth System Sciences, DOI 10.1007/978-3-642-25728-5_6,

analysis, (5) management strategy evaluation, (6) monitoring, as well as (7) ecosystem and performance evaluation.

In this chapter we focus on a necessary prerequisite of an IEA, i.e. an evaluation of the state and historical development of an ecosystem. This approach, also called the "aggregate approach" (Hall and Mainprize 2004), provides important information on the functioning of the ecosystem, baseline information necessary for a full IEA cycle. Methodologically, no common approach has been defined to describe the ecosystem state, diagnose sudden changes, and reveal responsible drivers and mechanisms. Here, we outline several methods that help investigating the ecosystem in a holistic way by considering all trophic levels as well as external drivers of change, such as fishing, eutrophication and climate. The statistical tools are robust, easy to use and have been successfully applied to a range of marine ecosystems, specifically on the Northeast U.S. continental shelf ecosystem (Link et al. 2002), the Scotian Shelf (Choi et al. 2005), the North Sea (Kenny et al. 2009) as well as for different sub-systems of the Baltic Sea (Möllmann et al. 2009; Lindegren et al. 2010; Diekmann and Möllmann 2010). In all studies large-scale reorganization of marine ecosystems, so called regime shifts, were detected.

In the following we will first provide some ecological background and present theories and definitions about patterns of ecosystem responses, resilience and regime shifts. Then, we will give technical advice on how to evaluate the state of an ecosystem and its historical development. We use the approach developed in e.g. Link et al. (2002), Choi et al. (2005), Kenny et al. (2009) and Möllmann et al. (2009) as the basis and step-by-step describe the procedure, from data considerations to the necessary statistical tools. Finally, we will present the analysis of the Central Baltic Sea as an example, and discuss the results with respect to the implementation within the EAM. Hands-on support for independently performing some of the described methods is included in the appendix.

6.2 Ecological Background: Theories and Observations

Marine ecosystems, like terrestrial ecosystems, are exposed to a number of external driving factors, which can be of physical and/or anthropogenic origin. Further internal mechanisms like natural fluctuations of species abundances influence the structure of the system, especially predator–prey interactions or niche overlaps. The effects of changes in the external drivers can be as complex as most ecosystems are and are neither easily detected nor described. Nevertheless, a number of theories exist how ecosystems react on changing environmental conditions and what are the consequences for a management goal to recover stressed ecosystems to an assumed pristine or at least more desirable state.

Scheffer et al. (2001) and Scheffer and Carpenter (2003) described three different patterns of ecosystem changes: When external conditions show a directional change, the ecosystem

1. May react in a smooth and continuous way, or
2. Shows a strong response when a critical level of the driver is reached, or
3. The so-called ecosystem response curve is folded backwards (S-shaped curve).

The latter means that two alternative stable ecosystem states exist, which are separated by an unstable equilibrium (middle part of the curve). This has two severe consequences. First, when a threshold level of the environmental driver is reached, the system shows an abrupt and catastrophic shift to a new state. Perturbations in addition to the change in external conditions may also induce the shift. Second, the system shows hysteresis. This means that it cannot be easily restored by returning to environmental conditions below this theoretical threshold level. The ecosystem rather remains in this alternative stable state until a significantly lower level or better conditions of the environmental driver (or combination of variables) is reached.

The ecological concept of two or more alternative stable states already emerged in ecological studies in the 1960s–1970s (Holling 1961; Rosenzweig 1971; May 1972; Holling 1973). Holling (1973) demonstrated the existence of multiple stability domains or multiple "basins of attraction" and how they relate to ecological processes and stochastic events. A useful measure of resilience is the size of the basins of attraction, or, in other words, the amount of disturbance a system can take before a shift takes place to a new stability region. If the resilience of a system is low, we can assume that the basin of attraction is eroded, the attraction to return to a certain state is reduced (Fig. 6.1), and even moderate perturbations may cause the system to shift.

The current definition of resilience goes beyond the initial term, now called the "Engineering Resilience", which can be interpreted as return time after disturbance (Holling 1996). It is rather defined as the capacity of a system to absorb disturbance

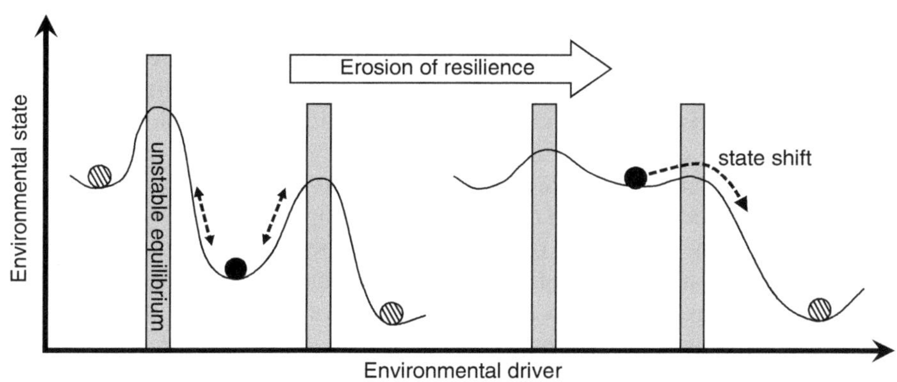

Fig. 6.1 Illustration of how an environmental driver can affect the resilience of an ecosystem with multiple stable states or multiple "basins of attraction" (valleys). The *black balls* represent the current state of the ecosystem and the *dashed balls* alternative states. With changing external conditions the resilience of the systems is reduced and the size of the basin of attraction gets eroded. Thus even moderate perturbations (*arrows*) may cause the system to shift

and re-organise while undergoing a change and still retain essentially the same function, structure, identity and feedbacks (Walker et al. 2004). Therefore, the relevant focus is not on constancy but on variability (Folke 2006). Nowadays, the concepts of renewal, regeneration and re-organisation are preferred over the term recovery (Bellwood et al. 2004); and similarly it seems to be more appropriate to use the word "regimes" instead of the terms "stable state" or "equilibrium", because they imply that dynamics are excluded (Scheffer and Carpenter 2003).

Ecosystem resilience is possibly negatively affected by environmental change and overexploitation. In this case a system is more vulnerable to disturbances and is at risk to shift to a new state. Such a "regime shift" is commonly defined as changes between contrasting persisting states of any complex system. More specifically, in marine ecology we understand it as a substantial reorganisation of the ecosystem state that persists long enough (e.g. more than a decade) in a new quasi-equilibrium state following changes that were primarily triggered by large-scale climate-induced variations (de Young et al. 2004). Regime shifts are considered to be abrupt in the sense that the time-span during which the shift occurs is short relative to the duration of the different states (de Young et al. 2008). Such large-scale reorganizations in the structure and function of the biological components have been detected in terrestrial, freshwater and marine ecosystems (Scheffer et al. 2001; Scheffer and Carpenter 2003), and for the latter numerous examples from all over the world now exist (e.g. Hare and Mantua 2000; Beaugrand et al. 2008; Choi et al. 2005; Cury and Shannon 2004; Howard et al. 2008), including studies from the Baltic Sea (Diekmann and Möllmann 2010; Lindegren et al. 2010; Möllmann et al. 2009).

Although a regime shift is often demonstrated by the abrupt change of one ecosystem property, e.g. the change in abundance of an important fish resource like cod (*Gadus morhua*) in the Baltic Sea, it always involves a number of ecosystem components and also driving forces. Marine ecosystem regime shifts have been explained as a result of multiple causes, involving both abiotic and biotic processes (Lees et al. 2006). Abiotic factors include temperature and salinity changes, turbulence, advection and eutrophication. Biotic drivers include restructuring of foodwebs due to overfishing and internal populations dynamics. Generally, both categories of drivers include natural and anthropogenic contributions which may operate synergistically, making it difficult to disentangle the various influences (Nyström et al. 2008; de Young et al. 2008).

Furthermore, the importance of drivers may vary regionally. While in the North Pacific clear indications exist for a climate-driven bottom-up change involving all trophic levels (Hare and Mantua 2000), the analysis of Frank et al. (2005) from the Scotian Shelf showed the indirect effects of fishing, because the decline of large predators like cod cascaded down the food web down to the level of nutrients. Strong multi-level trophic cascades have been also reported for enclosed systems, such as the Black Sea (Daskalov et al. 2007) and the Baltic Sea (Casini et al. 2008), the latter being additionally an example for a regime shift caused synergistically by climate and overfishing (Möllmann et al. 2009).

6.3 Guidelines for Investigating Ecosystem State and its Historical Development

Recently a series of studies were published analysing the status and temporal development of large marine ecosystems (LMEs) including the occurrence and timing of regime shifts (e.g. Beaugrand and Reid 2003; Conversi et al. 2010; Friedland and Hare 2007; Hare and Mantua 2000; Kenny et al. 2009; Litzow 2006; Mantua 2004; Molinero et al. 2008; Petersen et al. 2008; Solow and Beet 2005). However, no standardized approach was used to diagnose multivariate sudden changes and reveal the responsible drivers and mechanisms of ecosystem reorganizations. Reasons for this are inconsistent data bases as well as the prevailing shift type, i.e. if it is rather sudden or gradual, and if the changes are approximately synchronous across ecosystem components. The frequently applied analytical techniques each have their own conceptual or statistical problems, and their results can potentially deviate. This makes it particularly difficult to perform cross-ecosystem comparisons crucially needed for future decision-support systems within the EAM (Murawski et al. 2009). It is thus necessary to define guidelines for the different steps of a multivariate analysis of the state and development of an ecosystem, i.e. variable selection, data pre-treatment and the analytical techniques, which should involve a combination of robust and easy-to-use statistical methods. Below we will give recommendations for the various analytical steps and outline a series of methodologies, which can be used to analyse changes in ecosystem structure and function and that potentially allow future comparisons between ecosystems.

6.3.1 Data Mining

Here, we do not refer to monitoring strategies and data acquisition in terms of practical sampling and sample analysis. In its present meaning data mining rather involves collecting existing data series, from physical and biological monitoring programmes, climate indices, to fish and fisheries assessment metrics. Several aspects need to be considered and the first step in every ecosystem analysis must be an inventory about what information and time series are readily available. Based on the best scientific knowledge the key ecosystem components and environmental variables must be identified and included in the data collection. Here, data must not necessarily be available on the same temporal or spatial scale. In some cases it will be more appropriate to investigate seasonal data and/or data collected at a sub-system level to identify the most important characteristics of the ecosystem. If data cannot be included in the further analysis e.g. due to data gaps (see criteria below) they can be still of value for characterising the ecosystem or disentangling mechanisms influencing the system.

6.3.2 Variable Selection

Variables selected for integrated analyses should be quality controlled and must be usually condensed to annual estimates. Preferably they originate from monitoring or survey programs, i.e. represent raw data, and should not be model estimates. However, this criterion will be not always applicable as e.g. many fish stock parameters come from fisheries assessment models. Sometimes temporal and spatial effects within a system need to be considered and separate analytical runs may be necessary.

Generally the following conditions need to be fulfilled:

- Annual estimates of values need to be available (if necessary by averaging over time and space).
- Key components and parameters of the ecosystem need to be considered, i.e. key species and/or taxonomic groups representing all trophic levels and functional groups, hydrographic parameters, climatic indices, nutrients, other anthropogenic pressure variables (e.g. fisheries mortality).
- Missing values and data gaps must be minimised; in the optimal case no missing values are encountered. Missing data are especially critical at the beginning or the end of the data series or if subsequent estimates are missing. Most multivariate methods cannot deal with missing values,[1] and in this case variables can only be included if reliable side information exists to e.g. interpolate in a reasonable way.
- Cross-correlations between variables representing the same functional group or hydrographic/climatological processes should be low, i.e. information on processes/mechanisms/ecosystem components should not be "duplicated".
- Ecosystem components as well as forcing variables need to be balanced in terms of their number, even if they cannot be exactly equalised. If this criterion is not met the outcome of the integrated ecosystem analysis will be dominated by the dynamics of a single component (e.g. fish) and will not represent the entire ecosystem.
- There is no general rule how long time series need to be. However, to convincingly describe ecosystem dynamics over time, a length of at least 20 years should be achieved.

6.3.2.1 Considering Inter-Annual Dynamics

When inter-annual and seasonal dynamics are expected to be of importance for the overall dynamics of the ecosystem, separate analyses can be performed for each

[1] Solutions exist to perform multivariate analyses with missing values in the original data matrix, but will not be further discussed here. These are e.g. listwise deletion, pairwise deletion, imputation or maximum likelihood estimation for incomplete data.

season (e.g. spring and summer) or season-specific variable estimates need to be included. In the optimal case separate analyses can be performed for each season, but this is usually hampered by the availability of data.

6.3.2.2 Considering Spatial Dynamics

The dynamics of a large marine ecosystem (LMEs) like the Baltic Sea may not be the same as the ones observed in sub-systems that show some integrity of their own. According to the consideration of inter-annual effects it might be therefore necessary to perform (1) separate sub-system analyses, or (2) include variable estimates from different geographical areas in the overall analysis. In the latter case justifications need to be given why certain areas are included whereas others will not be considered.

In order to produce variable estimates, many parameters taken from monitoring activities need to be averaged over large geographical scales. In this case the spatial distribution of data points need to be considered and should generally be representative for the respective area. Furthermore, the spatial coverage should not vary considerably over time. However, if no consistent sampling or monitoring is available and the respective variable is still rated as important (this applies e.g. to many benthic organisms) it may be still included if the temporal dynamics of that specific variable or variable group is considered to be representative for the entire system.

6.3.3 Pre-Treatment of Data

Prior to any statistical analysis, a thorough data exploration needs to take place, including the investigation of cross-correlations between variables and especially between ecosystems components (see above), auto-correlations within single time series and data distributions. There is no general rule for data treatment prior to performing integrated ecosystem analyses. Transformations or standardisation procedures depend on the statistical methods used (see below). However some general aspects and methods can be described:

Weighting: A weighting of the data is usually not necessary. In case of strongly "unbalanced" data sets (i.e. the number of variables between system components is not equivalent) and no possibility to modify the data selection, a weighting of variables (e.g. considering the variables twice) can help around methodological problems in the data analysis. However, data modifications of this kind are rarely encountered in the literature.

Transformations: Data transformations are often necessary to linearise relationships between variables and change their distribution form e.g. from a skewed distribution to an approximately normal distribution. A number of transformations are possible. The most frequently used are:

$$\text{Logarithm}: \quad x^* = \log(x+1) \text{ for ``right - skewed''/positively skewed distributions (for } x \geq 0) \tag{6.1}$$

$$\text{Root}: \quad x^* = \sqrt{x} \text{ or } x^* = \sqrt[4]{x} \text{ for ``right - skewed''/positively skewed distributions (for x} \geq 0) \tag{6.2}$$

$$\text{Potentiation}: \quad x^* = x^a \text{ with a>1, for ``left - skewed''/negatively skewed distributions} \tag{6.3}$$

$$\text{Arcus sinus}: \quad p^* = \arcsin\sqrt{p} \text{ for relative and percentage data} \tag{6.4}$$

A right-skewed distribution is typical for biological data. After a thorough data exploration a common rule of thumb can be to log- or root-transform most biological data. It is recommended to use the same degree of transformation within variable groups.

Standardisation: For many statistical methods a standardisation of variables may be necessary. It is obligatory when performing e.g. a PCA on a suite of variables with different units and scales. This z-transformation/normalisation is often done automatically by statistical programs in case a "normalised PCA" or a "PCA based on the correlation matrix" will be performed (see below).

$$\text{z - transformation}: \quad z = \frac{x - \overline{x}}{s_x} = \text{normalisation} \tag{6.5}$$

$$\text{Proportioning}: \quad p = \frac{x}{\sum_{i=1}^{n} x_i} \tag{6.6}$$

6.3.4 Methods for Ecosystem Status Assessments

Many different approaches exist to investigate the ecosystem state and its historical development. Generally, a combination of exploratory methods with inferential statistics should be used. Here we give recommendations for data visualisation and for a suite of mathematical/statistical techniques combining ordination with objective methods to identify abrupt changes in single data series and across variables.

Computational hands-on support for applying most of the described methods with freely available software is given in the appendix.

1. *Traffic light/shade plot approach:* This is not an analytical but a plotting tool, which is especially suitable to visualise trends and changes considering all variables in the integrated analysis (for examples see Kenny et al. 2009; Link et al. 2002; Möllmann et al. 2009). Various authors use different approaches but most often values of each single variable using raw data are separated into quantiles (e.g. quintiles) and to each quantile a specific colour is assigned (e.g. green for low variable values, red for high variable values and colours in-between). To visualise the trends it is helpful to sort the variables, e.g. according to (1) their loadings on the first axis of a subsequently performed PCA, or (2) their standardised variable averages over the first three to five data points of the time series. Especially the sorting according to the respective PC1 loading results in a grouping of variables with similar temporal trends (due to cross-correlations between variables).
2. *Ordination:* Multivariate ordination techniques are useful data analysis tools for ecologists and can be also used to investigate time series data. Ordination plots object-points along one or several axes that represent an ordered relationship. Usually, the aim is to reduce the dimensions and represent the multivariate data in a bivariate diagram that represent a large fraction of the variability in the data matrix. Several methods exist, and the choice of the ordination technique is dependent on the nature of the data.

   For multivariate time series in ecosystem analysis, applying *Principal Component Analysis (PCA)* is a simple but useful approach. PCA is limited to quantitative data and cannot deal with missing values. Linear relationships between variables are assumed, which is only partly applicable to ecosystem data. Thus, data transformations of some variables (see above) and standardisation of all variables is a prerequisite, the latter is equivalent to performing a PCA based on the correlation matrix (= "normalised PCA").

   PCA is a so-called Eigenvector technique and aims at decomposing the variability in the dataset into new theoretical variables (the Principal Components or PCs), which are linear composites of the original variables. The PCs are orthogonal, i.e. independent of each other, with the first PC explaining most of the variance (described by the Eigenvalues) in the dataset, the second PC the second most, and so on. Generally, few PCs already represent a large fraction of the variability in the dataset. In the graphical display the PCs form the new axes and accordingly the objects (here: years) get new coordinates (scores). When variables and objects are shown in the same graph it is called a biplot. Dependent on the representation as a distance or a correlation biplot (see Legendre and Legendre 1998 for details) the distance between objects in a low-dimensional space approximates the Euclidean and the Mahalanobis Distance, respectively. Simplified, the closer two objects are together, the more similar they are according to the behaviour of their variables. Correspondingly, the original variables can be visualised as vectors with the vector apices representing the loadings on the respective axis. In the correlation biplot the angles between vectors approximate the degree of correlation between variables with a small angle representing a high correlation, a 90°-angle shows no correlation, and a 180°-angle a negative correlation.

Further details about the method, the mathematical background, and its interpretation can be found in Rao (1964), Legendre and Legendre (1998) and various statistical textbooks. Generally the following aspects need to be considered when performing a PCA:

- The number of variables between ecosystem components must be balanced.
- Cross-correlations within variable groups should be minimal, i.e. the same process or dynamic should not be described by more than one variable.
- Transformation of selected variables is necessary to linearise relationships.
- Missing data need to be minimised, and in case data gaps cannot be avoided they need to be filled according to best knowledge, e.g. taking temporal trends into account.
- PCA based on the correlation matrix should be performed (in this case, the data are normalised/z-transformed automatically).
- The correlation biplot should be plotted and consecutive years can be linked with each other for a better visualisation (distance between data points/years approximates the Mahalanobis Distance).
- Analysis can be repeated for abiotic forcing variables (e.g. climatic, physical parameters, nutrients, fishing pressure) and biotic response variables.

A couple of other ordination techniques exist and only two more should be shortly mentioned here:

Correspondence Analysis (CA) is sometimes also called "weighted PCA with χ^2-metric" and is computationally similar to the previously described PCA. Ter Braak (1985) showed that the χ^2-distance, used in CA, is a good approximation for species with a unimodal response curve to environmental gradients. However, it is only appropriate if the underlying gradients are strong and it has the preconditions that all variables need to show the same physical dimension and all values must be equal to or larger than zero. Thus, it is commonly used in community analyses, but is not applicable if different types of variables should be combined in one analysis. In contrast to PCA and CA, where the analysis is based on the Correlation/ Covariance matrix or the matrix of χ^2-Distances, any distance/similarity measure can be used in *Multidimensional Scaling (MDS)*. Thus, it is especially famous among community ecologists who prefer non-metric similarity coefficients being able to cope with the "double zero" problem (Field et al. 1982). Due to its flexibility this technique can be used with any kind of data to show (dis-) similarities between objects or variables. However, only two or three dimensions can be visualised, the axes have no numerical meaning and they are consequently not independent of each other. Thus, it is extremely difficult to derive any relationships to external gradients and driving forces from the ordination diagram.

3. *Discontinuity Analysis* tries to identify abrupt changes in data series either in space or time, and methods exist that are applicable to single or multivariate data sets.

Chronological Clustering is closely related to the commonly known Cluster Analysis, with the additional constraint that it can only group sequential samples (spatially or temporally) (Legendre et al. 1985). The basis for each clustering technique is the calculation of a similarity matrix. Thus, the first step is the selection of the appropriate similarity coefficient (Legendre and Legendre 1998). To make the results interpretable in relation to e.g. a previously performed PCA, the following procedure should be followed:

- Variables need to be standardised/normalised.
- The Euclidean Distance Matrix needs to be calculated (other similarity/distance coefficients can be used in case data subsets are analysed).
- "Proportional-link linkage algorithm" should be used to agglomerate/group samples → "non-hierarchical solution" to Cluster Analysis.

Technically Chronological Clustering is based on a permutation test, for which the significance level α and the connectedness level needs to be defined (Legendre et al. 1985) (recommendation: $\alpha = 0.01$, connectedness level $= 0.5$).

STARS is based on a sequential t-test that can signal a possibility of a regime shift in real time (Rodionov 2004): The method does not need an a priori hypothesis about the timing of the shift and has a good performance at the end of the time series, which means that a shift can be detected relatively early. In addition to its possibility to detect shifts in the mean level of fluctuations, it can also identify shifts in the variance, the latter algorithm being similar to that for the mean, but based on a sequential F-test. The code for the method is written in Visual Basic for Application (Excel) and can be downloaded from www.beringclimate.noaa.gov/regimes/.

The STARS method can be used for data of all kinds (raw data, anomalies, etc.) and is applicable to practically an unlimited number of variables. The program can additionally handle "red noise" time series. However, due to the limited length of most ecosystem time series a correction of autocorrelation is often not recommended (Rodionov 2006). An exception is if single time-series are analysed, which are known to have a high degree of autocorrelation, e.g. fish spawning stock biomass estimates (SSB).

The following parameters need to be defined before starting the analysis (for details see Rodionov 2004):

- Significance level α (e.g. $= 0.05$),
- Cut-off length (e.g. $= 8$ years),
- Huber's weight parameter (e.g. $= 2$).

For Integrated Ecosystem Analyses STARS can be applied (1) to all single variables, (2) to all variables or variable groups summing up the resulting Regime Shift Indices (RSI) for each point in time (bar plots of cumulative RSIs), and (3) to ordination scores resulting from previously applied dimension reduction techniques (e.g. PCs) and thus indicating overall ecosystem shifts as illustrated by e.g. the first PC-axes.

4. *Canonical Analyses*: In Canonical Analyses, variables are divided into response and explanatory variables, and both data matrices are directly compared to each other (direct gradient analysis). It is the appropriate method if *a priori* hypotheses exist about relevant factors and one is not interested in the general structure of the data (like in indirect gradient analyses, e.g. PCA). A variety of methods exist, and their application is partly dependent on the length of the underlying gradient. For relatively short gradients the extension of PCA, Redundancy Analysis (RDA) is mostly appropriate, for longer gradients assuming unimodal responses Canonical Correspondence Analysis (CCA), an extension of CA, can be applied. Often the results are summarised in a triplot, from which the relationships between explanatory (drivers) and response variables can be deduced. Furthermore, the significance of the explanatory variables and the optimal combination of factors can be tested by Monte Carlo permutation tests. For further information see Legendre and Legendre (1998) and Zuur et al. (2007).
5. *Other time series methods:* A couple of other time series tools exist that can be used to extract trends and analyse relationships to forcing variables. Here, only two methods should be shortly characterised:

   MAFA (Min/Max Autocorrelation Factor Analysis) is a principal component analysis related technique that extracts common trends in a time series data set (Shapiro and Switzer 1989; Solow 1994). The axes obtained by MAFA are smooth curves, or trends, the first MAF being the most important trend underlying all the original time series, the second MAF the second most important trend, etc. To make the derived trends interpretable, it may be necessary to analyse groups of variables separately.

   DFA (Dynamic Factor Analysis) is a dimension-reduction technique used to detect common patterns in a set of time series, relates these to explanatory variables and investigates interactions (Zuur et al. 2003, 2007). The technique can cope with missing values, and non-stationarity. It is based on multivariate structural time series models.

6.4 Integrated Ecosystem Assessment of the Central Baltic Sea (1974–2007)

The Central Baltic Sea (CBS) comprises three deep basins: Bornholm Basin (BB), Gdansk Deep (GD), and Gotland Basin (GB) (ICES Subdivisions 25, 26, 27, and 28) with a maximum water depth in the western Gotland Deep of 459 m. A characteristic of the deep basins is a permanent halocline separating low-salinity surface from highly saline deep waters. As a result of the stratification, deep-water salinity and dissolved oxygen conditions can only be renewed by lateral advection, i.e. major Baltic inflows of North Sea waters. The frequency of these events strongly regulates the abiotic conditions for the local fauna. The latter is generally of low diversity, with only a few species dominating the upper trophic levels.

Hence, the fish community is dominated by cod (*Gadus morhua*), herring (*Clupea harengus*), and sprat (*Sprattus sprattus*); the zooplankton community can be characterised by *Pseudocalanus acuspes*, *Temora longicornis*, and *Acartia* spp.

Similar to other areas of the world ocean, the Baltic Sea region experienced pronounced changes in atmospheric forcing in the past decades. This forcing can be well described by the Baltic Sea Index (BSI), which is correlated with the North Atlantic Oscillation index (NAO; Lehmann et al. 2002). The BSI was generally negative during the 1970s and 1980s, but turned positive during the late 1980s (Fig. 6.2). The change from negative to positive values was associated with more frequent westerly winds, warmer winters, and eventually a warmer climate over the Baltic Sea area. In addition to the direct influence on temperature, climate also affects salinity conditions in the Baltic Sea. Increased precipitation along with increased westerly winds negatively affected surface salinity. Additionally, during the high BSI period since the late 1980s, only two major Baltic inflow events were recorded. These are identified by a large amount of highly saline water (>17 psu) across the Darss and Drodgen sills and are quantified by the inflow intensity parameter Q96 as a function of the duration of the inflow and the mean salinity of the inflowing water (Fischer and Matthäus 1996; Fig. 6.2, upper right). The low frequency of major inflow events to the Baltic since the 1980s, although

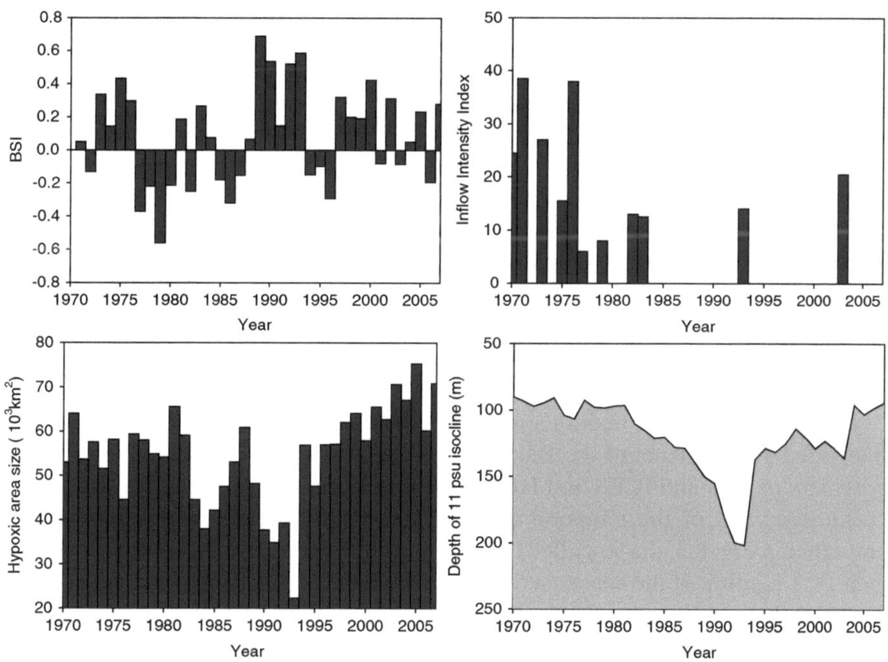

Fig. 6.2 Long-term hydroclimatic changes in the Baltic Sea indicated by the Baltic Sea Index (BSI), the inflow intensity parameter (based on Matthäus 2006; Feistel et al. 2008), the extend of hypoxic areas (≤ 2 mL L^{-1} O_2) in the Baltic (in km), and the depth of the 11-psu isocline in the Gotland Basin (in m) (from *upper left* to *lower right*)

unpredictable to date, has been hypothesized to be related to the high NAO, respectively BSI period (Hänninen et al. 2000). Increased run-off leading to sea level variations may have additionally hindered major inflow events (Matthäus and Schinke 1999). During the long "stagnation periods" since the early 1980s, virtually no high-saline, oxygen-rich waters were transported into the eastern basins, thus resulting in largely unfavourable conditions for marine species in deeper water and near-bottom layers. Although the size of hypoxic areas (i.e. ≤ 2 mL L^{-1} O_2) in the Baltic Sea decreased considerably during and after the strong inflow event in 1993, it increased nearly continuously since then and reached its maximum extent of about 75,000 km in 2005 (Fig. 6.2, bottom left) (Hansson et al. 2009). Deep-water salinity conditions, demonstrated by the depth of the 11-psu isoline in the Gotland Basin, also degraded since the early 1980s until the major inflow in 1993, when the halocline was located at a depth of approximately 200 m. Afterwards, conditions improved considerably, and more recently, the halocline was found at a depth of about 100 m (Fig. 6.2, bottom right).

6.4.1 Integrated Status Assessment

The ecosystem state and development of the Central Baltic Sea is described by abiotic and biotic time-series, providing information about climate and hydrography (17 variables), nutrients (8), phytoplankton (16), zooplankton (6), fish (9), and fisheries (3). All 59 data series were compiled to one estimate per year and covered the period 1974–2007.

Data were provided by various institutions, these are (in alphabetical order): Baltic Environmental Database of the Baltic Nest Institute (BED), Finnish Environment Institute (SYKE, former Finnish Institute of Marine Research, FIMR), Data Centre of the International Council for the Exploration of the Sea (ICES), Latvian Fish Resources Agency (LATFRA, now BIOR – Institute of Food Safety, Animal Health and Environment), Leibniz Institute of Marine Sciences (IFM-GEOMAR), Leibniz Institute for Baltic Sea Research (IOW), Swedish Meteorological and Hydrological Institute (SMHI).

Further, this study is mainly a result of the ICES/HELCOM Working Group on Integrated Assessments of the Baltic Sea (WGIAB) and we thank all participants, providers of data and ICES and HELCOM for their support (Table 6.1).

An overview of the temporal changes of all central Baltic Sea time-series is presented in Fig. 6.3, the so-called traffic light plot. Variables are sorted according to their PC1 loading of the subsequently performed PCA. This results in an order with variables that are linearly correlated to each other standing closely together. Thus, a pattern with variables at the bottom showing an increasing trend over time (green to red), with the highest values in the recent 15 years, to variables at the top demonstrating the opposite trend (red to green) with the highest values in the late 1970s to early 1980s is generated. The first group of variables comprises e.g. sprat, the copepods *Acartia* spp. and *Temora longicornis*, dinoflagellates, and temperature

Table 6.1 List of 59 variables used in the integrated status assessment of the Central Baltic Sea

Variable	Unit	Area	Abbreviation
Maximum ice cover	km^2	Baltic	MaxIce
Baltic Sea Index (Dec–Feb)	–	Baltic	BSI
Depth of 11-psu isoline	m	GB	11 psu_GBAnn
Sea surface temperature (Spr.)	°C	BB	SST_BB_Spr
Sea surface temperature (Sum.)	°C	BB	SST_BB_Sum
Sea surface temperature (Spr.)	°C	GB	SST_GB_Spr
Sea surface temperature (Sum.)	°C	GB	SST_GB_Sum
Midwater temperature (40–60 m) (Spr.)	°C	BB	T_BB_60_spr
Midwater temperature (40–60 m) (Sum.)	°C	BB	T_BB_60_sum
Midwater temperature (40–60 m) (Spr.)	°C	GB	T_GB_60_spr
Midwater temperature (40–60 m) (Sum.)	°C	GB	T_GB_60_sum
Sea surface salinity (Spr.)	psu	BB	SSS_BB
Sea surface salinity (Spr.)	psu	GB	SSS_GB
Halocline salinity (70–90 m) (Spr.)	psu	BB	S90_BB
Halocline salinity (80–100 m) (Spr.)	psu	GB	S100_GB
Deepwater oxygen (Spr.)	$mL\ L^{-1}$	BB	O2_BB
Deepwater oxygen (Spr.)	$mL\ L^{-1}$	GB	O2_GB
Dissolved inorganic nitrogen (surface) (Win.)	$\mu mol\ L^{-1}$	BB	DIN_BB_10_win
Dissolved inorganic phosphorus (surface) (Win.)	$\mu mol\ L^{-1}$	BB	DIP_BB_10_win
Dissolved inorganic nitrogen (surface) (Win.)	$\mu mol\ L^{-1}$	GB	DIN_GB_10_win
Dissolved inorganic phosphorus (surface) (Win.)	$\mu mol\ L^{-1}$	GB	DIP_GB_10_win
Dissolved inorganic nitrogen (deep water) (Sum.)	$\mu mol\ L^{-1}$	BB	DIN_BB_90_sum
Dissolved inorganic phosphorus (deep water) (Sum.)	$\mu mol\ L^{-1}$	BB	DIP_BB_90_sum
Dissolved inorganic nitrogen (deep water) (Sum.)	$\mu mol\ L^{-1}$	GB	DIN_GB_220_sum
Dissolved inorganic phosphorus (deep water) (Sum.)	$\mu mol\ L^{-1}$	GB	DIP_GB_220_sum
Chlorophyll a (Spr.)	$mg\ m^{-3}$	BB	Chla_BBSpr
Chlorophyll a (Sum.)	$mg\ m^{-3}$	BB	Chla_BBSum
Chlorophyll a (Spr.)	$mg\ m^{-3}$	GB	Chla_GBSpr
Chlorophyll a (Sum.)	$mg\ m^{-3}$	GB	Chla_GBSum
Diatomeae (Spr.)	$mg\ m^{-3}$	BB	dia_BB_spr
Dinoflagellata (Spr.)	$mg\ m^{-3}$	BB	dino_BB_spr
Cyanobacteria (Spr.)	$mg\ m^{-3}$	BB	cyano_BB_spr
Diatomeae (Sum.)	$mg\ m^{-3}$	BB	dia_BB_Sum
Dinoflagellata (Sum.)	$mg\ m^{-3}$	BB	dino_BB_sum
Cyanobacteria (Sum.)	$mg\ m^{-3}$	BB	cyano_BB_sum
Diatomeae (Spr.)	$mg\ m^{-3}$	GB	dia_GB_spr
Dinoflagellatea (Spr.)	$mg\ m^{-3}$	GB	dino_GB_spr
Cyanobacteria (Spr.)	$mg\ m^{-3}$	GB	cyano_GB_spr
Diatomeae (Sum.)	$mg\ m^{-3}$	GB	dia_GB_sum
Dinoflagellata (Sum.)	$mg\ m^{-3}$	GB	dino_GB_sum
Cyanobacteria (Sum.)	$mg\ m^{-3}$	GB	cyano_GB_sum
Acartia spp. (Spr.)	$mg\ m^{-3}$	GB	Acartia_Spr
Acartia spp. (Sum.)	$mg\ m^{-3}$	GB	Acartia_Sum
Temora longicornis (Spr.)	$mg\ m^{-3}$	GB	Temora_Spr
Temora longicornis (Sum.)	$mg\ m^{-3}$	GB	Temora_Sum
Pseudocalanus acuspes (Spr.)	$mg\ m^{-3}$	GB	Pseudo_Spr

(continued)

Table 6.1 (continued)

Variable	Unit	Area	Abbreviation
Pseudocalanus acuspes (Sum.)	mg m^{-3}	GB	Pseudo_Sum
Cod spawning stock biomass	t	SD25–32	CODSSB
Cod recruitment	N (age 2)	SD25–32	CODR2
Cod weight	kg (age 3)	SD25–32	CODWC3
Cod fishing mortality	(age 4–7)	SD22–32	COD_F47
Sprat spawning stock biomass	t	SD22–32	SPRSSB
Sprat recruitment	N (age 1)	SD22–32	SPRR1
Sprat weight	kg (age 3)	SD22–32	SPRWC3
Sprat fishing mortality	(age 3–5)	SD22–32	SPR_F35
Herring spawning stock biomass	t	Baltic	HERSSB
Herring recruitment	N (age 1)	Baltic	HERR1
Herring weight	kg (age 3)	Baltic	HERWC3
Herring fishing mortality	(age 2–6)	Baltic	HER_F26

Information is given about the unit of the respective variable, the area for which the variables are valid, and the abbreviation used in the following figures (*GB* Gotland Basin, *BB* Bornholm Basin, *Win.* Winter, *Spr.* Spring, *Sum.* Summer)

measures. Decreasing values were found e.g. for cod and herring spawning-stock biomass and recruitment, *Pseudocalanus acuspes* biomass, salinity measures, and the magnitude of the maximum ice extent. Variables with less clear temporal trends are found in the centre of the plot, some of them demonstrating relatively low values in the 1970s/1980s, high values between 1988 and 1993, and again low values thereafter. This group consists mainly of indicator time-series related to nutrients and phytoplankton variables; for some of them, measurements in the 1970s were missing, which might slightly deteriorate the respective sorting procedure.

The standardized PCA of the 59 variables resulted in 27.0% and 14.1% explained variance on the first two principal components (PCs; Fig. 6.4). The two-dimensional ordination plot (PC1 vs. PC2, Fig. 6.4 right panel) represents an abstract graphical presentation of the ecosystem state. Years close to each other are more similar according to their underlying descriptors than years further apart. Linking consecutive years with a line results in a time trajectory, denoting the theoretical ecosystem development over time. It demonstrates that changes in ecosystem states occurred most likely in 1987/1988, when the distance between these consecutive years was very pronounced, especially on the first and thus most important PC. This shift was further confirmed by chronological clustering as well as by STARS, for the latter using only PC1 scores, illustrated by the black dots in Fig. 6.4 (left panel). Here, it becomes evident that year scores change rapidly from positive to negative values in the late 1980s and remain negative since then. The development of PC2 scores is characterized by a steady decrease until the early 1990s and a sharp increase thereafter. Step-changes can be observed in the early 1980s and early 1990s. However, only the first change was found to be statistically significant using chronological clustering and STARS, and was possibly caused by the sudden drop in herring and cod SSB and recruitment time-series. These events preceded the main shift (1987/1988) that was detectable as abrupt

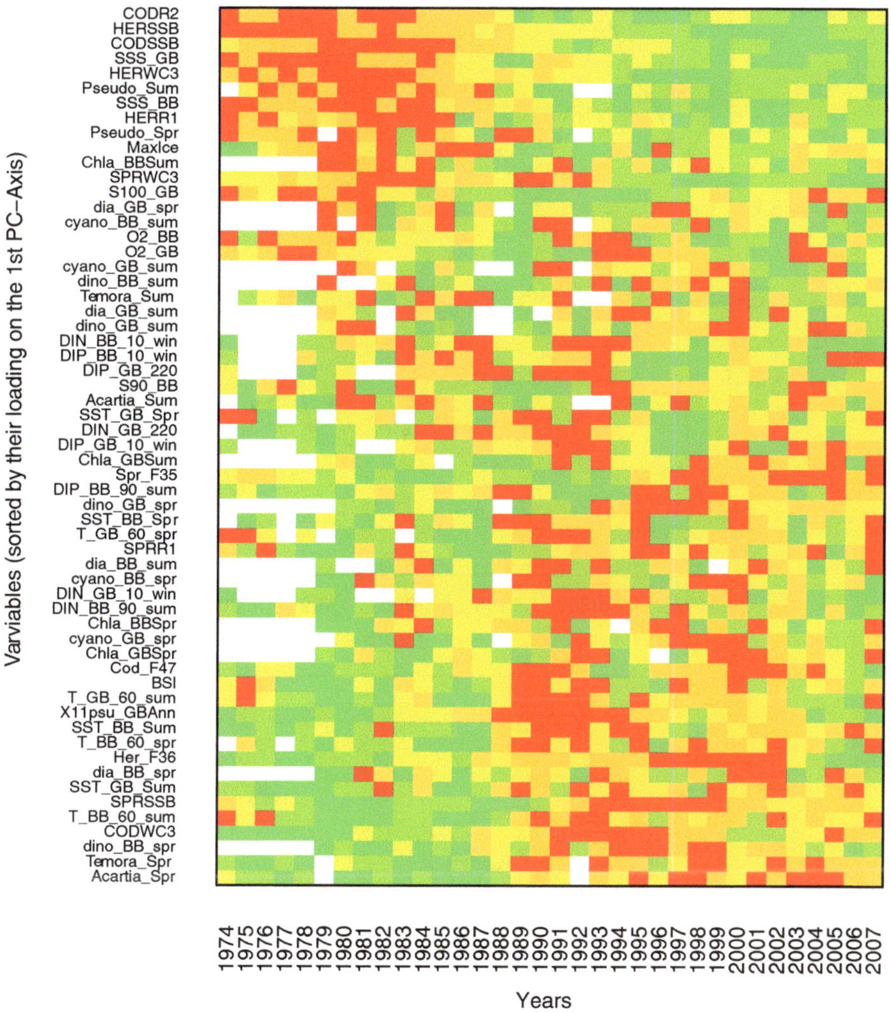

Fig. 6.3 Traffic-light plot of the temporal development of CBS time-series. Variables are transformed to quintiles, colour coded (*green* = low values; *red* = high values), and sorted in numerically descending order according to their loadings on the first principal component. Variable names are explained in Table 6.1

changes in the majority of all variables. The shift in the early 1990s was only found to be significant by STARS (1993). Temporarily it coincided with a major inflow event, and relates to changes in deepwater salinity, oxygen and nutrient concentrations. Since 1993, the distance between years was found to be rather small and they remained in the same quadrant of the plot (Fig. 6.4 right panel). This means that the ecosystem state of the central Baltic Sea in the past 15 years was relatively stable, and variable fluctuations were comparatively small.

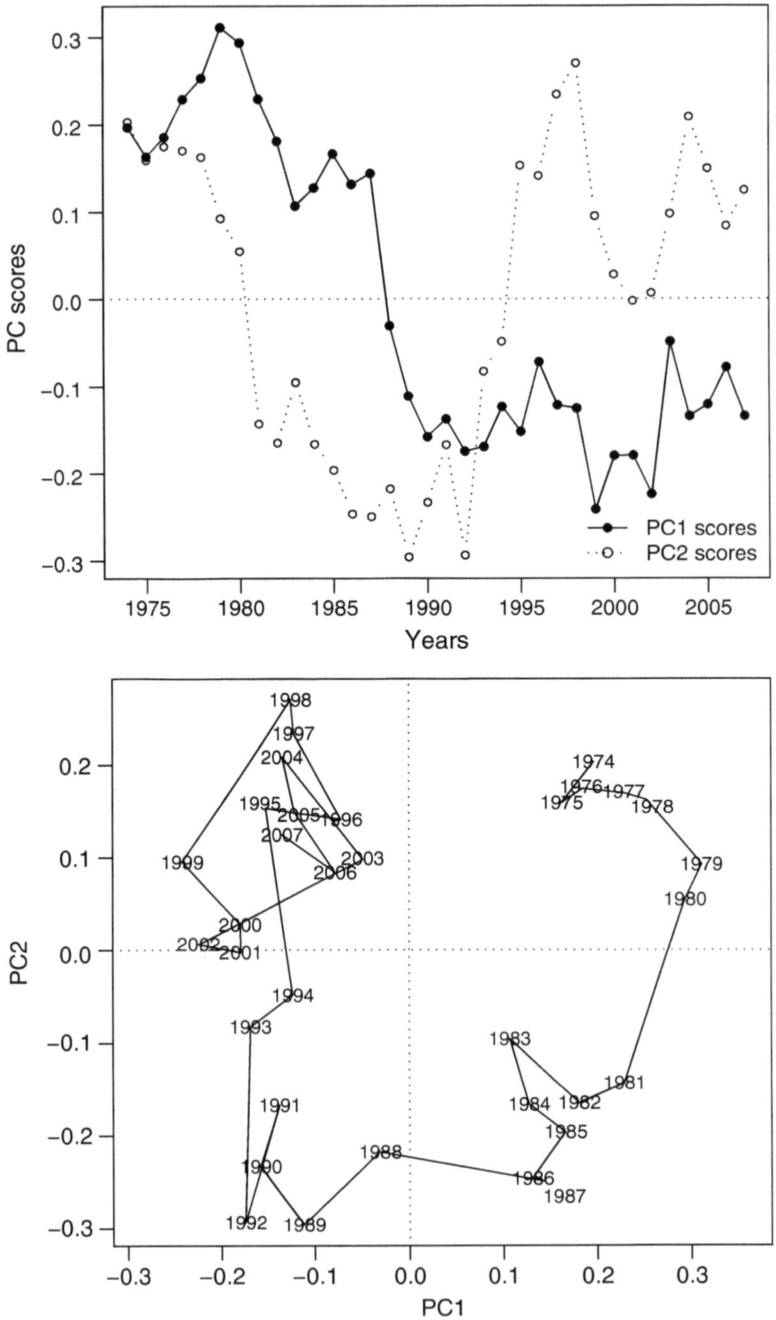

Fig. 6.4 Results of the standardized principal component analysis using all 59 variables assembled for the Central Baltic Sea showing PC1 (*black circles*) and PC2 scores (*white circles*) against time, and the time trajectory on the first factorial plane. PC1 = 27.0%, PC2 = 14.1% explained variance (from *left* to *right*)

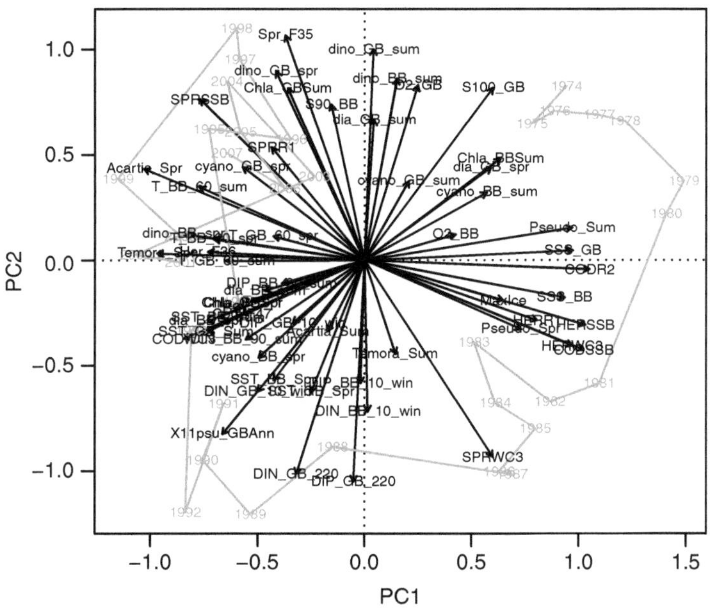

Fig. 6.5 Results of the standardized principal component analysis using all 59 variables assembled for the central Baltic Sea showing the variable loadings on the first factorial plane (for orientation: time trajectory from Fig. 6.4 in *light grey*). Variable names are explained in Table 6.1

The relative changes in the variables over time and in relation to the observed ecosystem shifts can be additionally derived from the factor loadings on the first two principal components (Fig. 6.5). In this biplot the year scores linked consecutively with each other are adumbrated in light grey. PC1 mainly reflects temperature (high negative loadings on PC1, indicating an increasing trend over time) and salinity (high positive loadings on PC1, indicating a decreasing trend over time). Highest negative PC1 loadings of biotic time-series were found for species known to have profited from the recent warming, e.g. sprat (Köster et al. 2003), *Acartia* spp., and *T. longicornis* (Möllmann et al. 2003) as well as Bornholm Basin dinoflagellates (Wasmund et al. 1998). In contrast, species which have suffered from the decrease in salinity, e.g. cod (Köster et al. 2005), *P. acuspes* (Möllmann et al. 2003), and herring (Möllmann et al. 2005), are negatively correlated with the previous group. Another factor that obviously contributed to the decline in the cod and herring stocks was high fishing pressure (represented as the fishing mortality coefficient F). F-values for both species have negative loadings on PC1, whereas the biomass development of both stocks is negatively correlated with fishing pressure and load positively on PC1. PC2 mainly reflects changes which occurred in deep water, i.e. during the long stagnation period until 1993, which decreased deep-water salinity and dissolved oxygen saturation (high positive loadings on PC2). In contrast, deep-water nutrients increased in this period (high negative loadings on PC2). After the reversal of conditions following the 1993 inflow, the same deep-water trends were observed until the recent inflow in 2003.

Generally, the pronounced change in the late 1980s seems to be driven by an increase in temperature as a result of the change in atmospheric forcing reflected by increasing values of the BSI time-series. It affected all ecosystem components and pushed the system into a new state. The smaller shift in 1993–1994 was mainly the result of deep-water-related processes, changes in nutrient loadings (decrease in 1993), and increases in salinity and dissolved oxygen concentrations following the 1993 inflow event. It is thus more a shift in abiotic conditions rather than a change in the whole ecosystem.

6.4.2 Resumée

The analysis presented above describes a regime shift in the pelagic ecosystem of the Central Baltic Sea predominantly during the late 1980s. It is quasi-synchronous with similar events detected especially in many North Pacific and North Atlantic marine ecosystems (e.g. Hare and Mantua 2000; Link et al. 2002; Beaugrand 2004; Choi et al. 2005; Weijerman et al. 2005). Further, multiple trophic levels were affected: The two regimes encompassing the years 1974–1987 (not taking the small change in the early 1980s into account, which might be statistically influenced by the many data gaps at the beginning of the time series) and 1994–2007 were characterised by the opposite dominance of key fish and zooplankton species, i.e. cod and *Pseudocalanus acuspes* in contrast to sprat and *Acartia* spp. (MacKenzie et al. 2007; Möllmann et al. 2008). The intermediate period from 1988 to 1993 showed a high variability across variables. Extreme values were partly observed for forcing variables such as low salinity and oxygen conditions, high temperatures and nutrient levels as well as high cod fishing pressure, which probably contributed to shifting the biotic part of the ecosystem into a new state (Möllmann et al. 2009). The changes in the abiotic environment occurred in a sequence of events, accumulating during the "transition period" (Möllmann et al. 2009).

Figure 6.6 (modified from Möllmann et al. 2009) summarizes the changes in the Central Baltic ecosystem in a conceptual diagram. It is based on the results of two separate PCAs of biotic response and abiotic driving variables, both not presented in this book. Characteristic species and physical factors shown in the diagram had a high temporal variability and were strongly correlated to the first principal axis of the respective analysis. In the diagram, two different indices were defined: (1) the ecosystem state index being an idealised curve of PC1-scores resulting from the analysis of biotic variables, and (2) the external forcing index being the equivalent for abiotic driving variables. The ecosystem change occurred during the transition period when the most extreme abiotic conditions prevailed, i.e. lowest salinity and oxygen, but highest temperature and nutrient levels. The formerly cod dominated ecosystem with *Pseudocalanus acuspes* as important copepod species in the food web then began to switch to a sprat-dominated system, with *Acartia* spp. as prevailing copepod taxon. After the transition period, the external forcing index returned approximately to its original state, while the ecosystem state index did not.

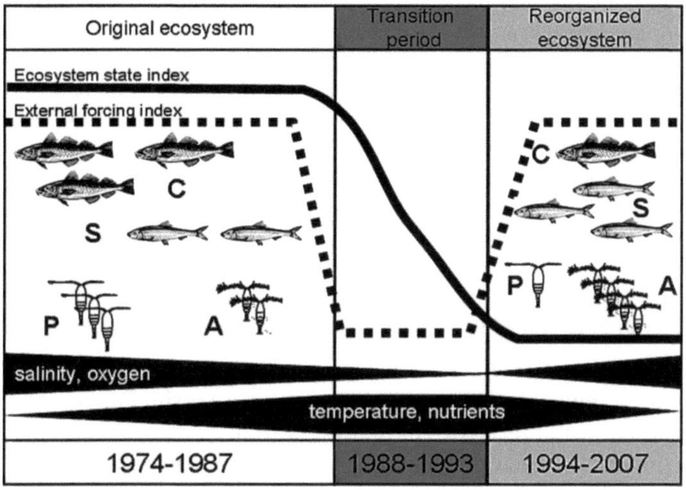

Fig. 6.6 Conceptual diagram displaying the changes in the Baltic Sea ecosystem; abbreviations are C for cod, S for sprat, P for *Pseudocalanus acuspes* and A for *Acartia* spp. (Modified from Möllmann et al. 2009). For further explanations, see the text above

This indicates the existence of hysteresis in the Central Baltic Sea ecosystem and characterizes the observed changes as a discontinuous regime shift (Scheffer and Carpenter 2003; Collie et al. 2004).

6.5 Concluding Remarks

The approach we outlined in this chapter is a first, but crucially important step towards implementing the ecosystem approach to management in marine environments. It provides knowledge on (1) the state and development of main ecosystem components, (2) the holistic behaviour of the ecosystem and (3) how the foodweb is responding to the mutual and interacting effects of external pressures such as climate, eutrophication and fisheries. The analyses described here use statistical methods, which have their limits and drawbacks. When indicated, complimentary methods should be used to test hypothesis about prevailing processes causing ecosystem changes. However, the presented analyses can lead to a significantly better understanding of changes in ecosystem structure and function with a special emphasis on catastrophic reorganizations such as ecosystem regime shifts.

The proposed approach forms the basis for establishing a full IEA as a concept for ecosystem-based management (Levin et al. 2009; Tallis et al. 2010). Specifically, it contributes to step 2) defining indicators and measures, step 3) setting thresholds, step 6) monitoring, as well as step 7) ecosystem and performance evaluation. To establish the full IEA process, the ecosystem analyses need to be further complemented by ecosystem modelling approaches which allow addressing the remaining IEA steps and providing the basis for a future risk analysis.

Acknowledgements The authors like to thank ICES, HELCOM and all participants of the ICES/HELCOM Working Group on Integrated Assessments of the Baltic Sea (WGIAB) for their support. We are further especially indebted to all people and institutions providing data for performing the status assessment of the Baltic Sea. Special thanks go also to the organisers and host of the NMA Summer School on "Climate Impacts on the Baltic Sea – From Science to Policy" for giving us the opportunity to write this article.

Appendix

Practical Guidelines for Performing Integrated Ecosystem Analyses

The analyses presented in the previous sections can be simulated with any appropriate dataset, i.e. with time series of multiple variables representing the ecosystem. Nowadays, indirect ordination techniques are implemented in nearly all commercial statistical software packages. For identifying breakpoints in single data series, different methods exist. Here we present methods and software that are all available as freeware and thus can be directly used by the reader. The analyses and plots in section 6.4 were mainly performed by the below described methods (see also Table 6.2). Whenever possible we used R and related packages as language for statistical computing (R Development Core Team 2007). The codes with short explanations will be described below. For users unfamiliar with R we refer to the website (www.R-project.org) and the manual "An introduction to R" (Venables et al. 2009). It is recommended to do the provided sample session first, before starting with any further analyses. Further, various books for beginners and advanced users exist and more will appear in near future. It is merely impossible to give an overview but a few recommendations can be found in Table 6.3.

Data Preparation

For the methods used in this chapter, data need to be organised in a $n * m$ data matrix, which means with n columns representing variables and m rows representing observations over time. Frequently, observations represent annual estimates for the specific variable (if necessary averaged over space and time). If seasonal specific values should be considered they need to be treated as separate variables or in separate analytical runs.

In extensive time series missing values are often encountered. These data gaps should be few and preferably should not occur in consecutive years. Data exploration techniques, single variable analyses (e.g. regime shift analysis of single time series (STARS)) and data illustration (e.g. annual anomalies, traffic light plot) should be based on the "raw" data ("*Originaldata*" and "*Originaldata2*" – see the R code described below). In contrast, for PCAs and consequently for the analysis of sudden changes in multivariate datasets, missing values need to be replaced and some variables need to be previously transformed. When performing a PCA many statistical programs exchange missing values with variable averages. However, if data show a long-term temporal trend, this might be obscured in the ordination plot.

Table 6.2 Software (mainly freely available) for performing various methods used in Integrated Ecosystem Assessments

Software	Methods	Authors	Availability	Web-link
STARS	Detection of shifts in both the mean level of fluctuations and the variances → Sequential t-test/F-test	S. N. Rodionov	Freeware, Add-in to Excel	www.beringclimate.noaa.gov/regimes/index.html
R-package (different from R CRAN)	Various complex multivariate and spatial analyses, includes chronological clustering	P. Casgrain, P. Legendre	Freeware, currently only available for MacOS	www.bio.umontreal.ca/Casgrain/en/labo/R/index.html
BRODGAR	Full statistical software. Includes various direct and indirect ordination methods, chronological clustering, DFA, MAFA, etc.	A. F. Zuur, E. N. Ieno	Commercial, stand-alone and with interface to R	www.highstat.com www.brodgar.com
R packages	Language and environment	R Development Core Team	Freeware	www.R-project.org
MASS		W. N. Venables, B. N. Ripley	Freeware, package in R	cran.r-project.org/web/packages/MASS/index.html
vegan	Direct and indirect ordination methods, diversity analysis for community analyses	J. Oksanen, R. Kindt, P. Legendre, B. O'Hara, G. L. Simpson, M. Henry, H. Stevens, and H. Wagner (2008).	Freeware, package in R	cran.r-project.org/web/packages/vegan/index.html
strucchange	Testing, monitoring and dating structural changes in (linear) regression models	A. Zeileis, F. Leisch, K. Hornik, C. Kleiber (2002).	Freeware, package in R	cran.r-project.org/web/packages/strucchange/index.html
rioja	Includes constrained clustering (→ chronological clustering)	S. Juggins (2009)	Freeware, package in R	cran.r-project.org/web/packages/rioja/index.html

Table 6.3 A short list of books introducing R and its use for statistical analyses (with one example for time series analysis)

Authors (publication year)	Title (publisher)
Zuur AF, Ieano EN, Meesters EHWG (2009)	A beginner's guide to R (Springer Science+Business Media)
Crawley MJ (2005)	Statistics: an introduction using R (John Wiley & Sons, Chichester)
Crawley MJ (2007)	The R book (John Wiley and Sons, Chichester)
Dalgaard P (2002)	Introductory statistics with R (Springer, New York)
Shumway RH and Stoffer DS (2006)	Time series analysis and its applications with R examples (Springer, New York)

Generally, we recommend exchanging data gaps with the average of the four nearest data points. This might be not possible if gaps exist at the beginning or the end of the time series or consecutive observations are missing. Therefore we didn't write a code for exchanging missing values but advise the readers to do it themselves, e.g. in their original data spreadsheet, considering their best knowledge about temporal trends and variability in the data. Following the exchange of missing values and with the previous knowledge about the distribution form of the data, transformations of some variables might be necessary. To improve linearity between variables and reduce the relationship between the variance and the mean a logarithmic ($log_{10}(x + 1)$, $ln(x + 1)$) or root transformation is recommended for right-skewed variables, which is often the case for biological observations. The complemented and partly transformed dataset ("*Data.transf*" – see the R code described below) can then be used for the various multivariate analyses presented below.

To be able to use R for the analyses, the three data sheets (here called "*datasheet.txt*", "*datasheet2.txt*", and "*datasheet_trans.txt*") with column and row labels need to be stored as separate tab-delimited text files. To be able to upload the data as described below you previously need to set the working directory to the folder containing your data (see "file" in the R toolbar). Other options are possible to import data (e.g. using a specific package for a direct import of data from Excel spreadsheets) but these will not be described here.

Analytical Steps

The following methods are not comprehensive and do not comprise all necessary or possible techniques for performing IEAs. As mentioned in Sect. 6.3 a thorough data exploration needs to take place to decide which transformations are necessary and which are the most appropriate tools to analyse the time series. We will not go into detail here but rather present examples how the time series can be graphically illustrated and how to accomplish some of the statistical analyses.

Temporal Anomalies

The temporal dynamics of variables can be presented in various ways. Here we show the R-code for a bar plot (Fig. 6.7) using deviations from the mean ($y' = x_{ij} - \bar{x}_i$)

```
# import data
Originaldata <- read.table("datasheet.txt", header = TRUE)
# check the data
str(Originaldata)

# data preparation: create a dataframe for the anomaly values
AnomData <- Originaldata
# save the dimensions (number of rows[1] and columns[2]) in a
# vector
n <- dim(AnomData)
# loop to convert actual values into anomaly values (incl.
# missing values in the average- and anomaly-calculation);
# the loop starts with the second column to exclude "Year"
# (no variable) and ends with the last column (defined by the
# second element of the vector n (n[2]) )
for (i in 2:n[2]) {
     AnomData[,i] <- AnomData[,i] - mean(AnomData[,i],
     na.rm = TRUE)}

# plot the data (here for the copepod species Acartia, in
# spring. $Acartia_Spr is the respective column title):
# to set the limit of the y-axis in the barplot (ylim)
y.low <- min(AnomData$Acartia_Spr, na.rm = TRUE) - 2
y.upp <- max(AnomData$Acartia_Spr, na.rm = TRUE) + 2
barplot(AnomData$Acartia_Spr, col = "black", space = 0.7,
  ylim = c(y.low, y.upp), axis.lty = "solid",
  names.arg = AnomData$Year, axisnames = TRUE,
  ylab = "Abundance anomalies (N/m³)",
  main = "Acartia in Spring")
box(lty = "solid", col = "black")
# creates a horizontal line at y = 0
abline(a = NULL, b = NULL, h = 0, v = NULL, reg = NULL,
       coef = NULL, untf = FALSE)
```

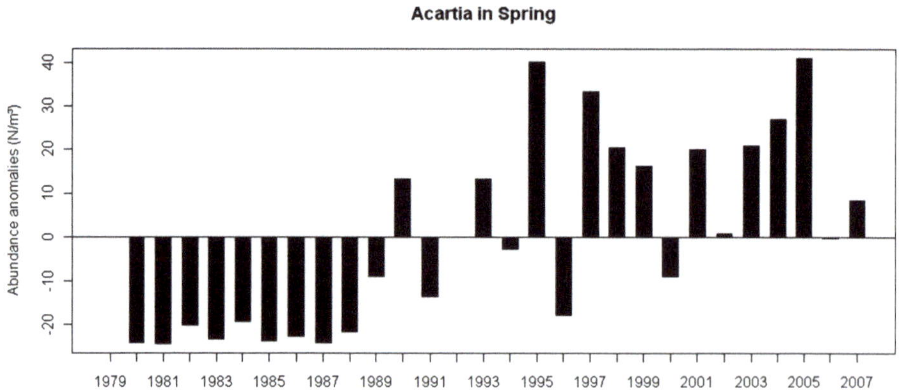

Fig. 6.7 A bar plot showing anomalies in spring abundances of *Acartia* in the period 1980–2007

Traffic Light Plot

This plotting tool is especially suitable to visualise trends and changes considering all variables in the integrated assessment. In the following example, each variable is divided into quintiles and to each quintile a specific colour is assigned (from green – low values, to red – high values). Variables are sorted in ascending order according to their standardised variable averages over the first five data points of the time series. An alternative is a sorting procedure according to the results of the subsequently performed PCA: Then variables are sorted with respect to their loadings on the first principal axis. However, the number of quantiles as well as the sorting procedure is not a convention and can be changed according to the personal needs (see also Fig. 6.3).

For this plot data, the original data file needs to be rearranged, in a way that the column with the years (ascending order) represents the ***last*** *column. The respective datasheet refers to "datasheet2.txt", or "Originaldata2" respectively.*

```
# data preparation before import: place the column with the
# years (ascending order) at the end!
# Save data as a text file.

# import data
Originaldata2 <- read.table("datasheet2.txt", header = TRUE)
# check the data
str(Originaldata2)
```

```
# data preparation: save the dimensions (number of rows[1]
# and columns[2] ) in a vector
n <- dim(Originaldata2)
# to get names for the plot axes except the last column
# (year)
# the y-axis (without the last column (years) )
Variablenames <-names(Originaldata2[1:n[2]-1])
# the x-axis
Years <- as.vector(Originaldata2[,n[2]])

# generates a matrix containing the quintile values for each
# variable (excluding the years by setting the column number
# and loop end to n[2] -1)
Quintiles <- matrix(nrow = 5, ncol = n[2]-1, byrow = FALSE)
for (i in 1:(n[2]-1)) {
     Quintiles[, i] <- quantile(Originaldata2[,i],
     probs = c(0.2, 0.4, 0.6, 0.8, 1.0), na.rm = TRUE)}

# generation of a dataframe where original values are
# converted into code 1 to 5 (depending on the quintile)
# using two loops: first create new dataframe based on the
# original dataframe
CodedData <- Originaldata2
# then convert the actual values into codes using a loop:
# the first (outer) loop repeats process for every variable,
# the second (inner) loop repeats process for every element
# within a variable
for(j in seq(along = CodedData[1:(n[2]-1)])) {
    for(i in seq(along = CodedData[[j]])) {
        if(is.na(CodedData[[j]][i]) == TRUE) {
        is.na(CodedData[[j]][i])<- TRUE} else
        if(CodedData[[j]][i] < Quintiles[1,j]) {
        CodedData[[j]][i]<-1} else
        if((CodedData[[j]][i] >= Quintiles[1,j]) &&
        (CodedData[[j]][i] < Quintiles[2,j])) {
        CodedData[[j]][i]<-2} else
        if((CodedData[[j]][i] >= Quintiles[2,j]) &&
        (CodedData[[j]][i] < Quintiles[3,j])) {
```

```
        CodedData[[j]][i]<-3} else
        if((CodedData[[j]][i] >= Quintiles[3,j]) &&
        (CodedData[[j]][i] < Quintiles[4,j])) {
        CodedData[[j]][i]<-4} else
        if(CodedData[[j]][i] >= Quintiles[4,j]) {
        CodedData[[j]][i]<-5} }}
# check if each column has the same counts of code values 1-5
CodedData

# plot preparation:
# to order variables in the plot in accordance to the
# standardised variable averages over the first five data
# points (excluding the years with n[2]-1 ):
# First create vector
Avg5year <- vector(length = n[2]-1)
# then fill the vector with the standardised variable
# averages over the first five data points (mean of first 5
# years - mean of full time series/ standard deviation of
# full time series)
for (i in 1:(n[2]-1)) {
     Avg5year[i] <- (mean(Originaldata2[c(1:5), i], na.rm =
     TRUE) -
     mean(Originaldata2[, i], na.rm = TRUE)) /
     sd(Originaldata2[, i], na.rm = TRUE) }
# finally, order the variable and create an index vector
OrdVar <- order(Avg5year)

# to sort coded dataframe in accordance to the 5-year
# standardised average, use the vector OrdVar as an index
CodedData.sorted <- CodedData[,c(1:n[2]-1)][OrdVar]

# convert dataframe into a matrix (necessary for an image
# plot)
CodedData.mat <- as.matrix(CodedData.sorted)

# image plot:
x <- c(1:n[1])
y <- c(1:n[2]-1)
op <- par(mar = c(4, 5, 3, 6), oma = c(0.5,2,0,0), xpd =
   TRUE)
image(x, y, z = CodedData.mat, zlim = c(1,5),
   col = c ("green", "yellowgreen", "yellow", "gold", "red"),
```

```
   axes = FALSE, xlab = "Years", ylab = "")
axis(1, at = seq(1, n[1], by = 1), tick = FALSE, labels =
   Years, cex.axis = 0.7, las = 3)
axis(2, at = seq(1, n[2]-1, by = 1), tick = FALSE, labels =
   Variablenames[OrdVar], cex.axis = 0.5, las = 1, padj = 1)
box()
title(main = "Traffic Light Plot", font.main = 4)
mtext("Variables (sorted by the 5-year standardised
   average)", outer = TRUE, side = 2)
# if you have a lower version than R2.10, remove the border
# argument
legend(x = max(x)+1, y = max(y), legend = c("< 0.2 quintile",
   "< 0.4 quintile", "< 0.6 quintile", "< 0.8 quintile",
   "> 0.8 quintile", "missing value"),
   fill = c("green", "yellowgreen", "yellow", "gold", "red",
   "white"),
   border = c("green", "yellowgreen", "yellow", "gold",
   "red", "grey"),
   cex = 0.6, bty = "n")
par(op)
```

For the following analyses and plots, we recommend to replace missing values by four-year averages. Further, some variables need to be previously transformed, e.g. by logarithmic transformations. The respective datasheet refers to "datasheet_trans.txt", or "Data.transf" respectively.

Principal Component Analysis (PCA)

Based on the correlation matrix ("normalised PCA"). This analysis can be performed for the full dataset, including abiotic forcing and biological response variables, as well as for data subsets, e.g. separately for biotic and abiotic variables. The code describes how to perform the analysis, get the output in terms of Eigenvalues, explained variance, scores and loadings, and a couple of plotting tools how to present the results graphically (Fig. 6.4, 6.5, 6.8).

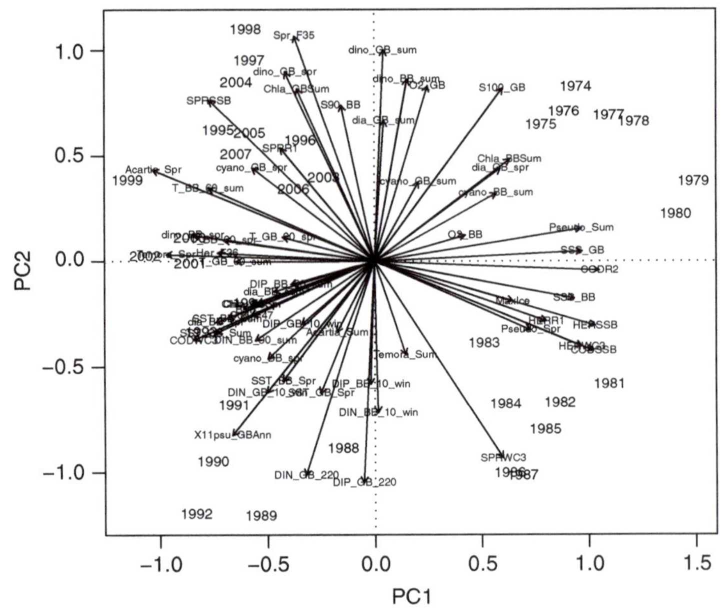

Fig. 6.8 A correlation biplot of a standardised Principal Component Analysis (PCA) displaying the loadings and scores, labelled here with the year, on the first two PC axes

```
# import data
Data.transf  <-  read.table("datasheet_trans.txt",  header  =
   TRUE)
# check the data
str(Data.transf)
# save the dimensions (number of rows[1] and columns[2]) in a
# vector
n <- dim(Data.transf)
# load package
library(vegan)

# PC analysis:
# exclude the years in PCA by using not the first column; to
# use the correlation matrix set the scale = TRUE)
PCA <- rda(Data.transf[, c(2:n[2])], scale = TRUE)
# to see the structure of the PCA result
str(PCA)
```

```
# output:
# to obtain the Eigenvalues
PCA.eigen <- PCA$CA$eig
PCA.eigen
# the summary gives the Eigenvalues, explained variance,
# cumulative variance, loadings (species scores) and scores
# (site scores, which are scaled by a constant given in the
# output, here 6.375328)
PCA.sum <- summary(PCA)
PCA.sum

# plots:

# create first a character vector containing the years for
# the score labels
years.char <- paste(Data.transf$Year, sep = "")

# biplot(with loadings and scores); to have more control of
# the plot, plotting items will be added separately. Variable
# names and scores will be not displayed by the setting:
# type = c("points","none"),
# the choices argument determines the PC axes (1:2 chooses
# PC1/PC2), the argument scaling = 3 scales species and sites
# symmetrically by the square root of the eigenvalues
biplot(PCA, choices = 1:2, scaling = 3,
   display = c("sites", "species"), type= c("points","none"),
   col = c("darkgreen"))
# add the variable names using same scaling value
text(PCA, display = "species", col = "darkgreen", cex = 0.5,
   scaling = 3)
# add now the scores with the year label using the same
# scaling value
text(PCA, display = "sites", labels = years.char, cex = 0.7,
   scaling = 3)

biplot(PCA, choices = 1:2, scaling = 3, display = c("sites",
   "species"), type= c("points","none"),col = c("darkgreen"),
   main = "Biplot with time trajectory")
# add now the scores with the year label using the same
# scaling value
text(PCA, display = "sites", labels = years.char, cex = 0.7,
   col = "grey", scaling = 3)
```

 # biplot with loadings and time trajectory in the back

```
biplot(PCA, choices = 1:2, scaling = 3, display = c("sites",
   "species"), type= c("points","none"),col = c("darkgreen"),
   main = "Biplot with time trajectory")
# add now the scores with the year label using the same
# scaling value
text(PCA, display = "sites", labels = years.char, cex = 0.7,
   col = "grey", scaling = 3)
# for the lines the coordinates of the scaled PC1/2 scores
# have to be obtained
scores.sc <- scores(PCA, scaling = 3)
scores.sc
# to connect the sites in an ascending order of the year
lines(x = scores.sc$sites[, 1][order(Data.transf$Year)],
   y = scores.sc$sites[, 2][order(Data.transf$Year)],
   lty = 1, col = "grey")
# add the variable names using same scaling value
text(PCA, display = "species", col = "darkgreen", cex = 0.5,
   scaling = 3)
```

For an example see Fig. 6.5.

 # Time trajectory plot with the unscaled scores

```
# in the plot, scores and loadings are not displayed (by
# setting: type = c("none","none"))
biplot(PCA, choices = 1:2, scaling = 0, display = c("sites",
   "species"), type= "n", main = "Time trajectory plot")
text(PCA, display = "sites", labels = years.char, cex = 0.7,
   scaling = 0)
# for the lines the coordinates of the unscaled PC1/2 scores
# have to be obtained
scores.unsc <- scores(PCA, scaling = 0)
# connect the sites in an ascending order of the year
lines(x = scores.unsc$sites[, 1][order(Data.transf$Year)],
   y = scores.unsc$sites[, 2][order(Data.transf$Year)],
   lty = 1, col = "blue")
```

For an example see Fig. 6.4, right panel.

 # Time series plot of PC1 and PC2 scores using results of the rda function

```
# get first unscaled PC1/PC2 site scores
scores.unsc <- scores(PCA, scaling = 0)
# get the min/max value for the limit of the x- and y-axis
y.low <- min(scores.unsc$sites)
y.upp <- max(scores.unsc$sites)
x.low <- min(Data.transf$Year)
x.upp <- max(Data.transf$Year)
plot(Data.transf$Year, scores.unsc$sites[, 1], pch = 16,
   ylab = "PC scores", xlab = "Years", axes = FALSE,
   xlim = c(x.low, x.upp), ylim = c(y.low, y.upp), cex = 0.7,
   cex.lab = 0.9, main = "Time-trajectory of PC scores",
   cex.main = 0.9)
axis(2, cex.axis = 0.7, las = 1)
# adjust sequence and labels to your respective time series!
axis(1, at = seq(1980, 2005, by = 5), label = seq(1980, 2005,
   by = 5), cex.axis = 0.7)
box(col = "black")
abline(h = 0, lty = "dotted")
lines(Data.transf$Year, scores.unsc$sites[, 1],
   col = "black", lwd = 0.8)
points(Data.transf$Year, scores.unsc$sites[, 2],
   pch = 1, cex = 0.7, col = "black", type = "b", lty = 3)
legend("bottomright", legend = c("PC1 scores", "PC2 scores"),
   text.col = c("black", "black"), cex = 0.7,
   col = c("black", "black"), pch = c(16, 1), lty = c(1, 3),
   pt.cex = 0.7, bty = "n")
```

For an example see Fig. 6.4, left panel.

"Constrained Clustering": Identifying Breakpoints in Multivariate Data

Chronological Clustering as performed in Sect. 6.4 is not yet available as function in R. So far it is implemented in the freeware "R-package", which is however only available for MacOS, and in the commercial software "BRODGAR" (Table 6.2). A similar, though not fully complementary because hierarchical method, is programmed in the package "rioja" for R (Functions for the analysis of quaternary science data). As agglomeration method it is recommended to avoid single linkage algorithms like "conslink" but rather applying the other option "coniss" using incremental sum of squares (Grimm 1987). Originally, this algorithm was developed for merging only stratigraphically adjacent clusters. This approach is similar as the one used for chronological clustering, although the latter method is preferred because other parameters can be defined and the statistical significance of breakpoints can be assessed.

In order to use constrained clustering in combination with PCA, data should be treated accordingly, i.e. in terms of missing values, variable transformations, standardisation of variables, as well as for the respective distance measure. In normalised PCAs, data are z-transformed and the Euclidean distance between observations is preserved in the plot. Therefore, a Euclidean distance matrix represents the input data for the analysis. Finally, a broken stick model is used to determine the number of significant groups in the analysis: As long as the output value for a specific number of groups is larger than the broken stick value the groups are termed significantly different (Fig. 6.9).

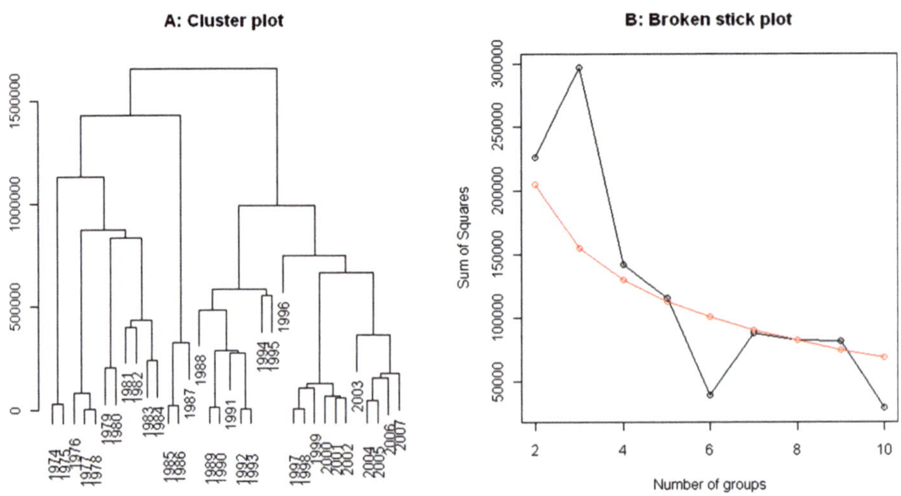

Fig. 6.9 (**a**) Cluster plot from the constrained clustering analysis using the "coniss" option, where row numbers were replaced by the year labels. (**b**) A plot of a broken stick model to determine the number of significant groups in the cluster analysis (here: five groups are significant because the values of the *jagged line* are above the critical value represented by the *smooth line*; but only for three groups the sum of squares are clearly above the value estimated from the broken stick model)

```
# import data
Data.stand <- read.table("datasheet_trans.txt", header =
   TRUE)
# check the data
str(Data.stand)

# save the dimensions (number of rows[1] and columns[2]) in a

# vector
n <- dim(Data.stand)

# load packages
library(vegan)
library(rioja)

# if data are not previously transformed (as in this case),
# the following code using the vegan package will standardise
# the data (necessary for the calculation of the Euclidean
# distance matrix from standardised data:
# decostand(x, method=standardize, MARGIN=2, na.rm=T)

# calculation of a Euclidean distance matrix from
# standardised data (stats package),
# excluding the first column with the years
Data.dist <- dist(Data.stand[, c(2:n[2])], "euclidean",
   diag = TRUE, upper = TRUE)


# Constrained hierarchical Clustering as described in "rioja"


  # package: Coniss = constrained incremental sum of squares
  # clustering
  CC <- chclust(Data.dist, method="coniss")

  # plot outcome, use the par function to display both plots
  # together:
  par(mfrow = c(1, 2))
  # cluster-plot with labels
  years.char <- paste(Data.stand$Year, sep = "")
  plot.chclust(CC,labels= years.char, cex = 0.7,
     main = "Coniss  - Cluster plot")
  # broken stick plot
  bstick.chclust(CC)
  title("Coniss - Broken stick plot")
```

Identification of Change-Points in Single Variables and a Combination of Variables

A series of methods exist, e.g. two-phase regression models (Solow 1987) or methods using least squares estimation (Lavielle and Moulines 2000). In R we refer to the package "strucchange", where tests for structural change in linear regression models from the generalized fluctuation test as well as from the F test framework are implemented. In Sect. 6.4 we used the freeware called "STARS", where the code is written in Visual Basic for Application (Excel) and can be downloaded from www.beringclimate.noaa.gov/regimes/. STARS is based on a sequential t-test that can signal a possibility of a regime shift in real time (Rodionov 2004). For the time series analysis described here, STARS can be used in three different ways:

1. Each single time series can be analysed for sudden changes separately;
2. All variables can be analysed at once. In the "Summary" spreadsheet Regime Shift Indices (RSI) are summed up for each point in time (bar plots of cumulative RSIs);
3. STARS can be applied to ordination scores of the first and, in case of a low amount of explained variance, additionally of the following PC-axes (resulting from the previously performed PC analysis).

The following parameters need to be defined before starting the analysis (for details see Rodionov 2004):

- Significance level α (e.g. = 0.05),
- Cut-off length (e.g. = 8 years),
- Huber's weight parameter (e.g. = 2).

It is recommended to use different parameter values e.g. for the cut-off length, and see how the output behaves. Depending on the overall length of the time series and the assumed length of the amplitude, other parameters than the ones mentioned above may be appropriate.

References

Beaugrand G (2004) The North Sea regime shift: evidence, causes, mechanisms and consequences. Prog Oceanogr 60:245–262

Beaugrand G, Reid PC (2003) Long-term changes in phytoplankton, zooplankton and salmon related to climate. Global Change Biol 9:801–817

Beaugrand G, Edwards M, Brander K, Luczak C, Ibanez F (2008) Causes and projections of abrupt climate-driven ecosystem shifts in the North Atlantic. Ecol Lett 11:1157–1168

Bellwood DR, Hughes TP, Folke C, Nyström M (2004) Confronting the coral reef crisis. Nature 429:827–833

Casini M, Lövgren J, Hjelm J, Cardinale M, Molinero J-C, Kornilovs G (2008) Multi-level trophic cascades in a heavily exploited open marine ecosystem. Proc R Soc B 275:1793–1801

Choi JS, Frank KT, Petrie BD, Legget WC (2005) Integrated ecosystem assessment of a large marine ecosystem: a case study of the devolution of the Eastern Scotian Shelf, Canada. Oceanogr Mar Biol Annu Rev 43:47–67
Collie JS, Richardson K, Steele JH (2004) Regime shifts: can theory illuminate the mechanisms? Prog Oceanogr 60:281–302
Conversi A, Fonda U, Peluso T, Molinero JC, Santojanni A, Edwards M (2010) The Mediterranean Sea regime shift at the end of the 1980s, and intriguing parallelisms with other European basins. PLoS One 5:e10633
Crowder LB, Hazen EL, Avissar N, Bjorkland R, Latanich C, Ogburn MB (2008) The impacts of fisheries on marine ecosystems and the transition to ecosystem-based management. Annu Rev Ecol Evol Syst 39:259–278
Cury P, Shannon L (2004) Regime shifts in upwelling ecosystems: observed changes and possible mechanisms in the northern and southern Benguela. Prog Oceanogr 60:223–243
Daskalov GM, Grishin AN, Rodionov S, Mihneva V (2007) Trophic cascades triggered by overfishing reveal possible mechanisms of ecosystem regime shifts. PNAS 104:10518–10523
de la Mare WK (2005) Marine ecosystem-based management as a hierarchical control system. Mar Policy 29:57–68
de Young B, Harris R, Alheit J, Beaugrand G, Mantua N, Shannon L (2004) Detecting regime shifts in the ocean: data considerations. Prog Oceanogr 60:143–164
de Young B, Barange M, Beaugrand G, Harris R, Perry RI, Scheffer M, Werner F (2008) Regime shifts in marine ecosystems: detection, prediction and management. Trends Ecol Evol 23:402–409
Diekmann R, Möllmann C (eds) (2010) Integrated ecosystem assessments of seven Baltic Sea areas covering the last three decades. ICES Coop Res Rep No. 302, 90 pp
Feistel R, Nausch G, Wasmund N (2008) State and evolution of the Baltic Sea, 1952–2005. A detailed 50-year survey of meteorology and climate, physics, chemistry, biology, and marine environment. Wiley-Interscience, Hoboken, p 703
Field JG, Clarke KR, Warwick RM (1982) A practical strategy for analysing multispecies distribution patterns. Mar Ecol Prog Ser 8:37–52
Fischer H, Matthäus W (1996) The importance of the Drodgen sill in the sound for major Baltic inflows. J Mar Syst 9:137–157
Folke C (2006) Resilience: the emergence of a perspective for social-ecological systems analysis. Global Environ Change 16:253–267
Frank KT, Petrie BD, Choi JS, Legget WC (2005) Trophic cascades in a formerly cod-dominated ecosystem. Science 308:1621–1623
Friedland KD, Hare JA (2007) Long-term trends and regime shifts in sea surface temperature on the continental shelf of the northeast United States. Cont Shelf Res 27:2313–2328
Grimm EC (1987) Coniss: a Fortran 77 program for stratigraphically constrained cluster analysis by the method of incremental sum of squares. Comput Geosci 13:13–35
Hall SJ, Mainprize B (2004) Towards ecosystem-based fisheries management. Fish Fish 5:1–20
Hänninen J, Vuorinen I, Hjelt P (2000) Climatic factors in the Atlantic control the oceanographic and ecological changes in the Baltic Sea. Limnol Oceanogr 45:703–710
Hansson M, Axe P, Andersson L (2009) Extent of anoxia and hypoxia in the Baltic Sea, 1960–2009. Swedish Meteor Hydrol Inst Rep, Dnr: Mo 214, 57 pp
Hare SR, Mantua NJ (2000) Empirical evidence for North Pacific regime shifts in 1977 and 1989. Prog Oceanogr 47:103–146
Holling CS (1961) Principles of insect predation. Annu Rev Entomol 6:163–182
Holling CS (1973) Resilience and stability of ecological systems. Annu Rev Ecol Syst 4:1–23
Holling CS (1996) Engineering resilience versus ecological resilience. In: Schulze P (ed) Engineering within ecological constraints. National Academy, Washington, DC, pp 31–44
Howard JAE, Jarre A, Clark AE, Moloney CL (2008) Application of the sequential t-test algorithm for analysing regime shifts to the southern Benguela ecosystem. Afr J Mar Sci 29:437–451
Juggins S (2009) rioja: analysis of quaternary science data, R package version 0.5–6 (www.cran.r-project.org/package=rioja)

Kenny AJ, Skjoldal HR, Engelhard GH, Kershaw PJ, Reid JB (2009) An integrated approach for assessing the relative significance of human pressures and environmental forcing on the status of large marine ecosystems. Prog Oceanogr 81:132–148

Köster FW, Möllmann C, Neuenfeldt S, Vinther M, St John MA, Tomkiewicz J, Voss R (2003) Fish stock development in the Central Baltic Sea (1976–2000) in relation to variability of the physical environment. ICES Mar Sci Symp 219:294–306

Köster FW, Möllmann C, Hinrichsen H-H, Tomkiewicz J, Wieland K, Kraus G, Voss R, Makarchouk A, MacKenzie BR, St. John MA (2005) Baltic cod recruitment – the impact of climate and species interaction. ICES J Mar Sci 62:1408–1425

Lavielle M, Moulines E (2000) Least-squares estimation of an unknown number of shifts in a time series. J Time Ser Anal 21:33–59

Lees K, Pitois S, Scott C, Frid C, Mackinson S (2006) Characterizing regime shifts in the marine environment. Fish Fish 7:104–127

Legendre P, Legendre L (1998) Numerical ecology, 2nd edn. Elsevier, Amsterdam

Legendre P, Dallot S, Legendre L (1985) Succession of species within a community: chronological clustering, with applications to marine and freshwater zooplankton. Am Nat 125:257–288

Lehmann A, Krauss W, Hinrichsen H-H (2002) Effects of remote and local atmospheric forcing on circulation and upwelling in the Baltic Sea. Tellus 54:299–316

Levin PS, Fogarty MJ, Murawski SA, Fluharty D (2009) Integrated ecosystem assessments: developing the scientific basis for ecosystem-based management of the ocean. PLoS Biol 7:e1000014

Lindegren M, Diekmann R, Möllmann C (2010) Regime shifts, resilience and recovery of a cod stock. Mar Ecol Prog Ser 402:239–253

Link JS, Brodziak JKT, Edwards SF, Overholtz WJ, Mountain D, Jossi JW, Smith TD, Fogarty MJ (2002) Marine ecosystem assessment in a fisheries management context. Can J Fish Aquat Sci 59:1429–1440

Litzow MA (2006) Climate regime shifts and community reorganization in the Gulf of Alaska: how do recent shifts compare with 1976/1977? ICES J Mar Sci 63:1386–1396

MacKenzie BR, Gislason H, Möllmann C, Köster FW (2007) Impact of 21st century climate change on the Baltic Sea fish community and fisheries. Global Change Biol 13:1–20

Mantua N (2004) Methods for detecting regime shifts in large marine ecosystems: a review with approaches applied to North Pacific data. Prog Oceanogr 60:165–182

Matthäus W (2006) The history of investigation of salt water inflows into the Baltic Sea – from the early beginning to recent results. Institut für Ostseeforschung Warnemünde. Mar Sci Rep 65:74 pp

Matthäus W, Schinke H (1999) The influence of river runoff on deep water conditions of the Baltic Sea. Hydrobiology 393:1–10

May RM (1972) Will a large complex ecosystem be stable? Nature 238:413–414

McLeod KL, Leslie HM (eds) (2009) Ecosystem-based management for the oceans. Island Press, Washington, DC

Molinero JC, Ibanez F, Souissi S, Buecher E, Dallot S, Nival P (2008) Climate control on the long-term anomalous changes of zooplankton communities in the Northwestern Mediterranean. Global Change Biol 14:11–26

Möllmann C, Köster FW, Kornilovs G, Sidrevics L (2003) Interannual variability in population dynamics of calanoid copepods in the Central Baltic Sea. ICES Mar Sci Symp 219:294–306

Möllmann C, Kornilovs G, Fetter M, Köster FW (2005) Climate, zooplankton and pelagic fish growth in the Central Baltic Sea. ICES J Mar Sci 62:1270–1280

Möllmann C, Müller-Karulis B, Kornilovs G, St John MA (2008) Effects of climate and overfishing on zooplankton dynamics and ecosystem structure: regime shifts, trophic cascade, and feedback coops in a simple ecosystem. ICES J Mar Sci 65:302–310

Möllmann C, Diekmann R, Müller-Karulis B, Kornilovs G, Plikshs M, Axe P (2009) Reorganization of a large marine ecosystem due to atmospheric and anthropogenic pressure: a discontinuous regime shift in the Central Baltic Sea. Global Change Biol 15:1377–1393

Murawski S, Steele JH, Taylor P, Fogarty MJ, Sissenwine MP, Ford M, Suchman C (2009) Why compare marine ecosystems? ICES J Mar Sci 67:1–9

Nyström M, Graham NAJ, Lokrantz F, Norström AV (2008) Capturing the cornerstones of coral reef resilience: linking theory to practice. Coral Reefs. doi:10.1007/s00338-008-0426-z
Oksanen J, Kindt R, Legendre P, O'Hara B, Simpson GL, Henry M, Stevens H, Wagner H (2008) Vegan: community ecology package. R package version 1.13–1. http://vegan.r-forge.r-project.org/
Petersen JK, Hansen JW, Laursen MB, Clausen P, Carstensen J, Conley DJ (2008) Regime shift in a coastal marine ecosystem. Ecol Appl 18:497–510
R Development Core Team (2007) R: a language and environment for statistical computing. R Foundation for Statistical Computing, Vienna. ISBN:3-900051-07-0. www.R-project.org
Rao CR (1964) The use and interpretation of principal component analysis in applied research. Sabjhyaá Ser A 26:329–358
Rodionov SN (2004) A sequential algorithm for testing climate regime shifts. Geophys Res Lett 31:L09204
Rodionov SN (2006) Use of prewhitening in climate regime shift detection. Geophys Res Lett 33: L12707
Rosenberg AA, McLeod KL (2005) Implementing ecosystem-based approaches to management for the conservation of ecosystem services. Mar Ecol Prog Ser 300:270–274
Rosenzweig ML (1971) Paradox of enrichment: destabilization of exploitation ecosystems in ecological time. Science 171:385–387
Scheffer M, Carpenter SR (2003) Catastrophic regime shifts in ecosystems: linking theory to observation. Trends Ecol Evol 18:648–656
Scheffer M, Carpenter S, Foley JA, Folke C, Walker B (2001) Catastrophic shifts in ecosystems. Nature 413:591–596
Shapiro DE, Switzer P (1989) Extracting time trends from multiple monitoring sites. Tech Rep No. 132. Department of Statistics, Stanford University, Stanford
Solow AR (1987) Testing for climate change: an application of the two-phase regression model. J Clim Appl Meteorol 26:1401–1405
Solow AR (1994) Detecting changes in the composition of a multispecies community. Biometrics 50:556–565
Solow AR, Beet A (2005) A test for a regime shift. Fish Oceanogr 14:236–240
Tallis H, Levin PS, Ruckelshaus M, Lester SE, McLeod KL, Fluharty D, Halpern BS (2010) The many faces of ecosystem-based management: making the process work today in real places. Mar Policy 34:340–348
ter Braak CJF (1985) Correspondence analysis of incidence and abundance data: properties in terms of a unimodal response model. Biometrics 41:859–873
Venables WN, Smith DM, The R Development Core Team (2009) An introduction to R. notes on R: a programming environment for data analysis and graphics. Version 2.10.1. http://cran.r-project.org/doc/manuals/R-intro.pdf
Walker B, Holling CS, Carpenter SR, Kinzig A (2004) Resilience, adaptability and transformability in social–ecological systems. Ecol Soc 9:5
Wasmund N, Nausch G, Matthäus W (1998) Phytoplankton spring blooms in the southern Baltic Sea – spatio-temporal development and long-term trends. J Plankton Res 20:1099–1117
Weijerman M, Lindeboom H, Zuur AF (2005) Regime shifts in marine ecosystems of the North and Wadden Sea. Mar Ecol Prog Ser 298:21–39
Zeileis A, Leisch F, Hornik K, Kleiber C (2002) Strucchange: an R package for testing for structural change in linear regression models. J Stat Softw 7:1–38, URL http://www.jstatsoft.org/v07/i02/
Zuur AF, Tuck ID, Bailey N (2003) Dynamic factor analysis to estimate common trends in fisheries time series. Can J Fish Aquat Sci 60:542–552
Zuur AF, Ieno EN, Smith GM (2007) Analysing ecological data. Springer, New York, p 672

Chapter 7
Sustainable Climate Science

Hans von Storch

7.1 The Social Institution "Science"

Doing science, creating new knowledge, in German: *Wissen schaffen*, is a social activity. As all social activities, it can be done sustainably. Or not.

The Brundtland Commission, convened by the United Nations (UN) in 1983, used the term "sustainability" in its 1987 report as 'development that meets the needs of the present without compromising the ability of future generations to meet their own needs.' Since then this term has been mostly used in the context of sustainable policy or sustainable use of natural resources – with a similar breadth as words like "ecological", which are now common in the language of advertising. It was no surprise that I was confronted with opposition when I used the word in a somewhat different context, namely in the context of science and research. As scientists, I am interested that my profession is done "sustainably". In Webster's dictionary (www.merriam-webster.com/dictionary/sustainability), a meaning of the word "to sustain" is "being a method of harvesting or using a resource so that the resource is not depleted or permanently damaged", for example "sustainable agriculture". Insert for the term "resource" the term "trust in the ability of science to generate reliable knowledge", and the meaning of "sustainable science" becomes clear. Or, more simply: A scientific institution or a professor is doing science sustainably, when some decades later the public and stakeholders assign to their former students the same authority for unraveling and explaining complex phenomena.

Studying classical Chinese language or the dramatic life of Abraham Lincoln is most likely done sustainably. Sustainability here means that the present science will

H. von Storch (✉)
Institute of Coastal Research, Helmholtz-Zentrum Geesthacht, Max-Planck-Str. 1, 21502 Geesthacht, Germany

KlimaCampus, Hamburg University, Hamburg, Germany
e-mail: hvonstorch@web.de

M. Reckermann et al. (eds.), *Climate Impacts on the Baltic Sea: From Science to Policy*, Springer Earth System Sciences, DOI 10.1007/978-3-642-25728-5_7,

have a bearing on future science, in an enriching way, but not in a limiting manner. The present science will not inhibit the legitimacy of future science. The public will be as excited about future knowledge as it is about present newly constructed insights.

Research on forest dieback in Germany may serve as an example at the other end of the spectrum. It was not done sustainably. The science of forest damages was in the 1980s heavily politicized, and used as support for a specific preconceived "good" policy of environmental protection. The resulting overselling and dramatization broke down in the 1990s, and news about adverse developments in German forests is now a hard sell in Germany. An observer (Keil 2004) wrote in 2004: "The damage for the scientists is enormous. Nobody believes them any longer."[1] Of course, the damage was not only limited to the forest researchers, but also to other environmental scientists and politicians as well.

And climate research? Often it is done sustainably, but sometimes not. Some institutions and some publicly visible scientists are known for simplifying and dramatizing statements of what one would expect from NGOs, e.g., "Coal-fired power plants are death factories" (Hansen 2009). A communication of drama is intended to "move", to initiate "action". The science is supposed to support a preconceived political agenda of something "good".

Overselling takes place in the triangle between policy, media and science. It goes with a risk (Weingart et al. 2000): The risk for policy-makers is that the goals set in this manner possibly cannot be achieved, the "loss of legitimacy due to taking on too much." The media primarily fear the "loss of public attention," due to concepts and conceptual fields becoming worn out. For science, the principal risk is the "loss of credibility due to the particular dynamic of the catastrophe metaphor", or other characteristic misleading concepts.

Exploiting short-term "advantages" in the public-political discourse by simplification and dramatization for driving a pre-conceived agenda helps generating attention and short-lived support for this agenda. But this attention and support can hardly be maintained for a long time as required in case of climate change policies. Attributing the hurricane Katrina to climate change made successfully headlines, and depicting global warming as an uninterrupted continuous upward trend made the understanding of the concept of global warming easier. But, later, we have to pay a prize. There were no more Hurricane-disasters like Katrina's since 2005, and warming is stagnating in the last years. Both facts are not surprising for the climate researcher. They are consistent with the scientific understanding of the phenomenon named "Global Warming". However they are at odds with the simplifying-dramatizing communication strategy and with the resulting medial construction.

The maximization of this short-term utility goes with a prize: The public will understand that it has been manipulated, and that it had not honestly been advised

[1] "Der Schaden für die Wissenschaftler ist enorm. Nun glaubt ihnen keiner mehr." op. cit.

by its publicly funded social institution "science". Admittedly, manipulated for something, which has been perceived by certain elites as "good" – but what is the principal difference to Greenpeace and Exxon in this respect? The dramatic decay of trust into the IPCC, following the illegal publication of e-mails at CRU in November 2009, the unacceptable sloppiness in preparing some statements, for instance about the future of Himalayan glaciers, in a report of the Fourth Assessment report of the IPCC, was not really surprising.

The effect is twofold. First, the public will no longer believe in the "story", or consider it merely entertainment – and people will effectively become sceptics. Certainly the contrary of the originally intended effect! Second, the public will be unable to distinguish the social institution science[2] from value-based NGOs – with the latter being considerably cheaper in delivering the same politically useful knowledge claims!

7.2 Conditions for a Sustainable Practise of Climate Science

Thus, for the sake of the credibility of our young PhD students, for maintaining the role of science as a useful social service and resource for society – we scientists have the obligation to act sustainably, and not to subordinate our credibility as auxiliary tool to a broader politically defined "good".

This obligation for acting sustainably requires:

1. *Admitting that scientific knowledge is uncertain and often in need of future revision. This admission does not mean that this lack of certainty would prevent the knowledge to be used. The public as well as decision makers are used to act under uncertainty.*

In case of climate science it has been estimated that about 10% of all scientists harbour some doubts whether it is really the emission of greenhouse gases which causes the present warming (Bray and von Storch 2007). On the other hand, about 80% subscribe to this explanation, which is a high percentage and may be taken as evidence for significantly good arguments, pointing towards the emissions as dominant driver of the ongoing changes and the perspective of future further warming. Among the sceptics, many are driven by secondary motives (Lahsen 2008), but some have purely scientific reservations. There is nothing bad about such scientific reservations.

2. *Understanding that the public discourse about climate change is driven by two different knowledge claims – a scientific and a media construction (von Storch* 2009a, b*)*

[2] Which, in the spirit of the US National Research Council is supposed to impartially sort out complex questions about the functioning and dynamics of our natural and social systems.

The *scientific* construct of human-made climate change is widely supported within the relevant scientific communities (Bray and von Storch 2007), and has been comprehensively formulated particularly thanks to the collective and consensual efforts of the UN climate council IPCC. According to this construction[3], humans are changing the global climate. Climate is the statistics of the weather. In almost all localities, at present and in the foreseeable future, the frequency distributions of the temperature are shifting to higher values and will continue to do so; sea level is rising; amounts of rainfall are changing. Some extremes such as heavy rainfall events will change. The driving force behind these alterations is above all the emission of greenhouse gases, in particular carbon dioxide and methane, into the atmosphere, where they interfere with the radiative balance of the Earth system.

Of course, there is no complete consensus in the scientific community, so that speaking of "the scientific construct" is a simplification, which is applied here for describing the contrast to the media construct, which is equally not just one construct but features many different variants.

This *media* construct may be summarized in this way: Climate is changing because of human activity. The weather is less reliable than it was before, the seasons more irregular, the storms more violent. Weather extremes are taking on catastrophic and previously unknown forms. The cause? Human greed and stupidity. The mechanism: the justice, or the revenge, of a nature that is striking back. For large chunks of the population, at least in Central and Northern Europe, the mechanism is obvious.[4] In old times, the adverse climatic developments were the just response to a God angered by human sins; this approach is also today, in particular in the US and possibly in the UK, sometimes invoked.[5]

[3] The term "construction" is not meant to belittle the scientific process or to hint to manipulation but merely to the fact that the knowledge has been obtained in a building process, exploiting different forms of knowledge; see also Müller and von Storch 2004: *Computer Modelling in Atmospheric and Oceanic Sciences – Building Knowledge*. Springer Verlag Berlin, Heidelberg, New York, 304 pp, ISN 1437-028X.

[4] On 14 August 2002, the reputable Swedish daily newspaper "Dagens Nyheter" wrote: "Naturen slår tilbaka våldsamt." (Nature strikes back violently), when reporting about a disastrous flooding in the Czech Republic.

[5] An example is provided by a former Chair of a IPCC Working Group who expressed his conviction that God would speak to the public through disasters (Welch 1995: Me and My God. *Sunday Telegraph*, 10.9.1995). Or, as it is put on the back cover of an alarmistic book "Our drowning world" (Milne 1989: *Our drowning world*, London: Prism Press): "...we shall be engulfed by the consequences of our greed and stupidity. Nearly two thirds of our world could disappear under polar ice cap water ... For this will be the inevitable outcome of industrialization, urbanization, overpopulation and the accompanying pollution." An enlightened variant is suggested by Lovelock in the framework of his Gaia hypothesis, when he speaks about "The revenge of Gaia – why Earth is fighting back..." (Lovelock 2006: *The revenge of Gaia – Why Earth is fighting back – and How we can still save humanity*. Penguin Group, London, 177 pp.).

3. *Accepting that consensual political conclusions about required or meaningful measures cannot be drawn from the understanding of the dynamics of climate change and of future perspectives alone (Pielke* 2007*).*

From the diagnosis that humans are changing climate, and that the time-integrated global emissions determine the magnitude of change, there is no immediate conclusion about the political implications to be drawn. Climate science is not "telling us what to do", but science is telling us- what the effect of which political conclusion would be. To this end, climate science should do research on the range of options, on their efficiency and costs, drawbacks and advantages. When results of this sort are available, the social decisions should be left to the democratic system. There should be a separation between scientific analysis and political decision making.

Interestingly, it was found for Northern Germany that people hardly have an idea about the regional or even local needs and options concerning required adaptation strategies to deal with the unavoidable part of future climate change (Ratter et al. 2009).

4. *Accepting that accurate language is needed – and that scientific terminology may conflict with every day language (e.g. Bray and von Storch* 2009*).*

An important example is the usage of the term "prediction" or "forecast". In climate we have not yet predictions, which would be a description of the most probable future development. We do not have such predictions, because we cannot predict future emissions and we have not yet developed methods to determine the present state sufficiently accurately. We have "scenarios" or "projections", which are descriptions of a series of possible futures. The latter are considerably less certain than the former, but useful nonetheless – if applied adequately.

For instance, a trend labeled as "statistically significant" does not necessarily extend into the future, or must be of anthropogenic origin. A counter example is the seasonal warming: The February–July temperature trend is significant, natural and does not extent into September.

5. *Explaining that reality is really complex and not simple. There are various examples when simplifications lead to incorrect and misleading conclusions.*

Sometimes the argument was brought forward that the warming of the atmosphere would necessarily bring about an intensification of mid-latitude storms – because of elevated water vapor levels. If the argument would be correct, we would not have winter but summer storms. And indeed, we cannot identify an intensification of "our storms" (e.g. Bärring and von Storch 2004) in Northern Europe in spite of a clear regional warming since 1873 (Cappelen 2008).

Interested economic quarters argue that the dramatic increase in damages to land-falling storms along the US coast would be evidence for intensifying hurricanes, but an alternative, better explanation is that economic value along the coasts has dramatically increased in recent decades (Pielke et al. 2008).

Mozart's song "Come, dear month of May, and make the trees flowering again" is no proof of late flowering of trees in Mozart's Vienna in the late eighteenth century, and thus cool conditions; the lines were written by a former major of Lübeck, 700 km north of Vienna (Böhm 2008). More generally, the assessment of data needs the expert knowledge to sort out issues of inhomogeneity (e.g. Peterson et al. 1998) (contamination of large-scale climate signals by other origins, such as the process of observing, processing etc.).

7.3 Science Communication and Service

Based on these general insights, concrete measures for a sustainable communication between science, the public, media, political and economic decision makers are needed.

On the **global** and **continental** scale, Working Group I of the Intergovernmental Panel of Climate Change IPCC provides reasonable communication, even if the IPCC suffers from some limitations. Examples are a partial politicization, the fact that frequently dominant authors rely on their own work, or the failure of Working Group 2 to accept the assessment of Working Group 1 about the dynamics and change of climate when assessing the impacts of climate.

On the **regional and local** scale, however, the IPCC is much less efficient, but different regional and local bodies provide services. An early excellent example was the Rossby Center at the Swedish Meteorological and Hydrological Institute in Norrköping. The Helmholtz Association has set up a number of Regional Climate Offices in different parts of Germany (Schipper et al. 2009); also a national Climate Service Center is beginning its operation in that country.

What needs to be done is (von Storch and Meinke 2008):

6. *The establishment of contacts for enabling a public-science dialogue. Here, the term "public" refers to the society at large, the media and political and economic decision makers.*

For the direction science-to-public, this implies that users of climate knowledge and information qualitatively and quantitatively understand scientific knowledge about climate, climate change and impacts – and do not need to rely on unreliable media reporting and claims-making by interested parties. This includes the understanding of uncertainty, limitation in spatial and temporal resolution, the character of scenarios, and the non-existence of prognoses as such (apart from a few exceptions so far), the role of natural variability, and the presence of ensemble spread.

For the direction public-to-science, the need is the understanding of concerns and questions in the public realm as well as among decision makers (Kempton et al. 1995); it includes also the analysis of the media construction of climate change.

The linear concepts of "informing", "teaching" or other types of "knowledge speaks to power/lay people" should be avoided, as they cannot provide the needed qualified feedback, nor can they deal with the ubiquitous alternative knowledge claims. They also fail to acknowledge the limitations of the scientific actors in understanding the social, cultural and political context.

7. *The determination of possible future climate changes needs worst-case scenarios*[6] *as well as best-case scenarios. The latter would describe the regional and local climate changes, which would have to be expected in case of successful, or even very successful international climate protection measures – and would thus represent "unavoidable" climate changes, which would require in any case adaptive responses.*

Additionally, knowledge about *options* to deal with expected adverse change is needed. Such options may include changes in urban planning (Gill et al. 2007), including the undoing of urban heat island effects, or different hydrodynamic regimes in estuaries (von Storch et al. 2008). Also, it needs to make clear that even very efficient local emission reductions do not provide shelter from the effect of global warming.

8. *The legitimate scientific knowledge about the ongoing climate change and its relationship to different drivers (urbanization, global change, land-use etc.) needs to be reviewed and assessed.*

Here, "legitimate" relates to publicly available publication originating from institutions operating with a "good scientific practice" standard. The idea should not be to describe "best" knowledge, as the "best"-claim is often a social construction, but consensus, including the consensus on disagreement, should be described. In a sense, such reports mimic the IPCC reports for specific regions.

Two such regional consensus reports have been prepared in Europe – one for the for the Baltic Sea region (BACC Author Team 2008); the other for the metropolitan region of Hamburg (von Storch and Claussen 2011). Both reports describe clear changes in regional temperatures and plant phenology, provide many regional climate change scenarios – but also a general lack of detailed knowledge about impacts and options.

9. *Finally, detailed data sets are needed, which after suitable quality control and homogenization describe regional and local climate change in past decades. Such data are needed to determine to what extent anthropogenic or other drivers (such as urbanization) have already had an effect on regional and local conditions (Bhend and von Storch* 2009*). They are also needed to determine present risks of extreme events.*

Such data sets may be constructed from various observations, local (e.g. from the CRU data sets, Jones and Moberg 2003) and satellite (e.g. from the Global

---

[6] For instance the work of the Dutch "Delta-Commissie" in 2008 on worst-case flooding and storm surge scenarios along the Dutch coast; see http://www.deltacommissie.com/en/advies

Precipitation Climatology Centre, Schneider et al. 2008), regional re-analyses or downscaled global re-analysis (Weisse et al. 2009).

This task is unfortunately often overseen, when the emphasis of the work is put mostly or even entirely on possible future developments. Indeed, the assessment of the recent past and of the ongoing change should be considered equally important than the design of scenarios of possible futures.

7.4 Epilogue

The present essay calls for a sustainable practice of climate science. Such a sustainability requires that science is not degenerating to a sort of politics and policymaking; instead, science is supposed to provide cold, dispassionate knowledge about the options of policymaking. To achieve this, "good science" is essential, which stands for openness, reproducibility, falsification, publication. And the acceptance that science, the pursuit of satisfying curiosity, is a part of our culture.

References

BACC Author Team (2008) Assessment of climate change in the Baltic Sea Basin. Springer, Berlin/Heidelberg, 473 pp. ISBN 978-3-540-72785

Bärring L, von Storch H (2004) Northern European storminess since about 1800. Geophys Res Lett 31:L20202. doi:10.1029/2004GL020441, 1–4

Bhend J, von Storch H (2009) Is greenhouse gas forcing a plausible explanation for the observed warming in the Baltic sea catchment area? Boreal Environ Res 14:81–88

Böhm R (2008) Heiße Luft – Reizwort Klimawandel. Edition Va Bene, Vienna

Bray D, von Storch H (2007) Climate scientists' perceptions of climate change science. GKSS-Report 11/2007, Geesthacht. http://coast.gkss.de/staff/storch/pdf/GKSS_2007_11.pdf

Bray D, von Storch H (2009) "Prediction" or "projection"? The nomenclature of climate science. Sci Commun 30:534–543. doi:10.1177/1075547009333698

Cappelen J (2008) Vejret som det gik i 2007 – i Danmark, i Nuuk på Grønland og i Tórshavn på Færøerne, Vejret 115

Gill SE, Handley JF, Ennos AR, Paulett S (2007) Adapting cities for climate change: the role of the green infrastructure. Built Environ 33:115–133

Hansen J (2009) Coal-fired power stations are death factories. Close them. The observer, Sunday 15 Feb 2009 quoted after http://www.guardian.co.uk/commentisfree/2009/feb/15/james-hansen-power-plants-coal. Accessed 30 Jan 2010, 00.00 UTC

Jones PD, Moberg A (2003) Hemispheric and large scale surface air temperature variations: an extensive revision and an update to 2001. J Climate 16(2):206–223

Keil G (2004) Chronik einer Panik. Ein Vierteljahrhundert Waldsterben – oder wie ein deutscher Mythos entstand, sich verfestigte und allmählich zerbröckelt. Beobachtungen aus dem Bundesforschungsministerium. DIE ZEIT 51/2004

Kempton W, Boster JS, Hartley JA (1995) Environmental values in American culture. MIT Press, Cambridge, MA/London, p 320. ISBN 0-262-11191-8

Lahsen M (2008) Experience of modernity in the greenhouse: a cultural analysis of a physicist "trio" supporting backlash against global warming. Global Environ Change 18:204–219

Lovelock J (2006) The revenge of Gaia – why earth is fighting back – and how we can still save humanity. Penguin Group, London, p 177
Milne A (1989) Our drowning world. Prism Press, London
Müller P, von Storch H (2004) Computer modelling in atmospheric and oceanic sciences – building knowledge. Springer, Berlin, p 304, 1437-028X
Peterson TC, Easterling DR, Karl TR, Groisman P, Nicholls N, Plummer N, Torok S, Auer I, Boehm R, Gullett D, Vincent L, Heino R, Tuomenvirta H, Mestre O, Szentimrey T, Saliner J, Førland E, Hanssen-Bauer I, Alexandersson H, Jones P, Parker D (1998) Homogeneity adjustments of in situ atmospheric climate data: a review. Int J Climatol 18:1493–1517
Pielke R Jr (2007) The honest broker. Cambridge University Press, Cambridge
Pielke RA Jr, Gratz J, Landsea CW, Collins D, Saunders M, Musulin R (2008) Normalized hurricane damages in the United States: 1900–2005. Nat Hazards Rev 9:29–42
Ratter B, Lange M, Sobiech C (2009) Heimat, Umwelt und Risiko an der deutschen Nordseeküste. Die Küstenregion aus Sicht der Bevölkerung. GKSS 2009/11. ISSN 0344-9629, pp 110
Schipper JW, Meinke I, Zacharias S, Treffeisen R, Kottmeier CH, von Storch H, Lemke P (2009) Regionale Helmholtz Klimabüros bilden bundesweites Netz. DMG Nachrichten 1-2009-10-12
Schneider U, Fuchs T, Meyer-Christoffer A, Rudolf B (2008) Global precipitation analysis products of the GPCC. Technical report, Global Precipitation Climatology Centre (GPCC), Deutscher Wetterdienst
von Storch H (2009a) Klimaforschung und Politikberatung – zwischen Bringeschuld und Postnormalität. Leviathan, Berliner Zeitschrift für Sozialwissenschaften 2009, 37:305–317. doi:10.1007/s11578-009-0015-8
von Storch H (2009b) Climate research and policy advice: scientific and cultural constructions of knowledge. Environ Sci Policy 12:741–747. http://dx.doi.org/10.1016/j.envsci.2009.04.008
von Storch H, Claussen M (eds) (2011) Klimabericht für die Metropolregion Hamburg. Springer, Berlin, p 300 (in German). ISBN 978-3-642-16034-9
von Storch H, Meinke I (2008) Regional climate offices and regional assessment reports needed. Nat Geosci 1(2):78. doi:10.1038/ngeo111
von Storch H, Gönnert G, Meine M (2008) Storm surges – an option for Hamburg, Germany, to mitigate expected future aggravation of risk. Environ Sci Policy 11:735–742. doi:10.1016/j.envsci.2008.08.003
Weingart P, Engels A, Pansegrau P (2000) Risks of communication: discourses on climate change in science, politics and the mass media. Public Underst Sci 9:261–283
Weisse R, von Storch H, Callies U, Chrastansky A, Feser F, Grabemann I, Günther H, Plüss A, Stoye T, Tellkamp J, Winterfeldt J, Woth K (2009) Regional meteo-marine reanalyses and climate change projections: results for Northern Europe and potentials for coastal and offshore applications. Bull Amer Meteor Soc 90:849–860. http://dx.doi.org/10.1175/2008BAMS2713.1
Welch F (1995) Me and my god. Sunday telegraph, 10.9.1995

Index

M. Reckermann et al. (eds.), *Climate Impacts on the Baltic Sea: From Science to Policy*, Springer Earth System Sciences, DOI 10.1007/978-3-642-25728-5,

MIX
Papier aus verantwortungsvollen Quellen
Paper from responsible sources
FSC® C105338

If you have any concerns about our products,
you can contact us on
ProductSafety@springernature.com

In case Publisher is established outside the EU,
the EU authorized representative is:
Springer Nature Customer Service Center GmbH
Europaplatz 3, 69115 Heidelberg, Germany

Printed by Libri Plureos GmbH
in Hamburg, Germany